The Star Finder Book

A Complete Guide to the Many Uses of the 2102-D Star Finder

Third Edition

David Burch

2102-D Star Finders available at
starpath.com

STARPATH

Seattle, Washington

Copyright © 1984, 1995, 2001, 2008, 2019 by David F. Burch

All rights reserved. No part of this book may be reproduced or transmitted in any form or by any means, electronic or mechanical, including photocopying, recording, or any information storage or retrieval system, without permission in writing from the author.

ISBN 978-0-914025-63-4

Published by

Starpath Publications

3050 NW 63rd Street, Seattle, WA 98107

www.starpathpublications.com

10 9 8 7 6 5 4 3 2 1

Preface to original 1984 edition

After teaching celestial navigation to over 3,000 students during the past 6 years, we have well learned the recurring questions. We especially appreciate questions and comments of former students after they navigate their first ocean crossing. Comments from new navigators are invaluable to the development of teaching methods and course materials. This booklet is one example.

Most discoveries of new navigators fall (with some shaking) into two categories: the good news and the bad news. The good news is their newly learned celestial navigation all works as it's supposed to. The sextant is not hard to use, even bouncing around in rough seas; and once the sextant sights are in hand they are confident they know where they are and how to get to where they want to go. The bad news that leads to most questions new navigators want answered before their next crossing is, skies are much more cloudy in the ocean than they thought they would be.

Under cloudy skies, the extreme value of accurate dead reckoning becomes clear very quickly. But this is not the subject at hand, and improving dead reckoning procedure is easily accommodated underway without special training. The problem with cloudy skies we cover here is of identifying unknown stars in isolated patches of clear sky. Such stars may be in view for a few minutes only, just long enough to get a sextant sight of them.

New navigators soon learn that these unknown stars may offer the only sights available for several days, and that this can occur much more often than was suspected. This is not just true in high-latitude oceans, famous for cloudy skies; it is a potential problem in all oceans including those of the tropics. It is, therefore, a basic skill of the navigator to know how to identify stars. Fortunately, this skill is easily mastered on land, long before it might be critical to a safe, efficient ocean passage.

In the past, we told students the best way to identify unknown stars is the 2102-D Star Finder, and we would spend an hour or so explaining its use. Then off they would go with this extent of preparation for the problem, whenever it should arise. We have since learned that this is not enough, and this booklet is intended to remedy this. This booklet explains the function and use of the Star Finder slowly and thoroughly, using detailed numerical examples and practice problems. In addition, the booklet discusses general features of star and planet choices and other practical procedures in celestial navigation.

Besides basic star identification, this booklet also elaborates on several other, more specialized, applications of the Star Finder—again, in response to questions from new navigators. These include its use in predicting heights and bearings of stars, moon, and planets for routine sights and picking optimum star-planet combinations. Navigators soon learn the value of the moon, Venus, or Jupiter for combination with star sights, since these three bodies can be seen during the brighter part of twilight when stars are faint but the horizon is still sharp.

We also demonstrate how the Star Finder can be used to answer some common questions in celestial navigation: Which days of the month offer a simultaneous sun-moon fix, what is the optimum time of day to do the sights, and how long must one wait between successive sun lines for a good running fix? These are fundamental, practical problems that are most easily solved using the Star Finder.

Another purpose of this booklet is to teach star identification to candidates for Merchant Marine licenses. Some examinations require star identification but do not require the candidate to know the details of celestial navigation by stars other than Polaris. The 2102-D Star Finder is an acceptable way to solve this part of the examination—it is also the easiest way, regardless of background in celestial navigation.

And finally, we would hope that this booklet might serve to introduce this Star Finder to interested stargazers outside the field of marine navigation. Anyone who knows (or cares to learn) how to use an Almanac, can use the Star Finder to identify all stars and planets, and predict their heights and bearings at any time. I suspect that if more people simply knew that the 2102-D Star Finder existed, more people would be using it, in more varied applications.

Preface to the 1995 printing

This printing remains the same as the first 1984 edition, except that a few typographic errors have been fixed and the name of the booklet has changed. Within the practice of celestial navigation, the main change has been the increased use of computers. They do star ID very conveniently, but—although the number of our students has increased to now over 13,000 and our own experience expanded some 30,000 miles—the conclusions of the early preface remain the same. The old plastic star finder will have a place onboard so long as small craft go to sea. It is the only convenient means of star ID that works when it's wet, and strangely enough, it is still more versatile than any software yet available.

Preface to the Second Edition

Some things have changed, some remain the same. We have now been fortunate to have had well over 20,000 classroom students at Starpath School of Navigation and thousands more in home study courses around the world, and my own ocean experience has grown to well beyond 60,000 miles underway. And, needless to say, technology has changed a lot. But the discussion of the First Edition Preface about the role of the Star Finder remains valid. The main computerized challenge to the device comes from our own StarPilot software programs for PCs and TI calculators, but they too are vulnerable to the elements underway, as all electronics are.

The most significant development since 1995 related to star and planet ID is the advent of several wonderful resources on the internet. Worth special mention is the US Naval Observatory site which computes heights and bearings of all the major bodies in the sky for any date, time, and location. One can now master the Star Finder for any location on earth by simply using the data they provide to check your results. See www.starpath.com/usno.

Preface to the Third edition

Numerical examples were carefully reviewed, which led to several corrections and clarifications. Nearly all the illustrations and images were improved, and we added new sections on improving accuracy from the template readings and how to solve great circle sailing. The Bibliography and other cited links were updated.

Contents

	Page
PREFACE	iii
LIST OF TABLES AND EXAMPLES	vi
Chapter 1. INTRODUCTION	1
Chapter 2. DESCRIPTION OF THE 2102-D STAR FINDER	3
2.1 The Original Navy Star Finder	7
2.2 British Star Finder, NP323	7
Chapter 3. BACKGROUND	8
3.1 Terminology and Celestial Motions	8
3.2 Brightness and Color of Stars and Planets	11
3.3 Tips on Planet Identification	15
3.4 Use of Moon and Planets	16
3.5 Time and Time Keeping in Navigation	18
Chapter 4. APPLICATION I: Star Identification after the Sight	22
4.1 Identifying Navigational Stars	22
4.2 Identifying Other Stars	25
4.3 Identifying Planets	27
Chapter 5. APPLICATION II: Predicting Heights and Bearings	30
5.1 Precomputing Stars for Routine Sights	31
5.2 Choosing Optimum Stars for Routine Sights	32
5.3 Precomputing Sun, Moon and Planets	35
5.4 Choosing Time Between Sun Lines for a Running Fix	36
5.5 Picking Optimum Star-Planet Combinations	40
Chapter 6. APPLICATION III: Sun-Moon and Sun-Venus Sights	48
6.1 Figuring Optimum Sight Times	49
6.2 A Precaution Doing Daylight Sights	55
6.3 Star Finder versus Alternative Methods	55
Chapter 7. Other Star Finder Applications	57
7.1 Emergency Steering with the Star Finder	57
7.2 Great Circle Sailing	58
BIBLIOGRAPHY	61
APPENDIX	62
Almanac Data	62
Improved Template Readings	67
Equation Summary	68

List of Numerical Examples

	Page
Example 3-1. Figuring the Brightness of Stars and Planets	15
Example 3-2. Finding the UTC and WT of Twilight.	20
Example 3-3. More on Finding UTC and WT	21
Example 4-1. Star ID: Navigational Stars	23
Example 4-2. Star ID: Non-navigational Stars	27
Example 4-3. Planet ID	28
Example 5-1. Precomputation for Star Sights	33
Example 5-2. Precomputing Sun, Moon and Planets	36
Example 5-3. Picking the Time between Sun Sights for a Running Fix	38
Example 5-4. Precomputing Star-Planet Combinations	43
Example 5-5. More on Star-Planet Combinations	44
Example 5-6. Precomputing Moon, Planet, Star Combinations	44
Example 6-1. Optimizing Sun-Moon and Sun-Venus Sights	51
Example 6-2. Picking the Time for Sun-Moon Sights	53

List of Tables

Table 2-1. Star Symbols on the White Disk	3
Table 3-1. Brightness versus Magnitude for Stars and Planets	10
Table 3-2. Navigational Stars that are Distinctly Reddish	13
Table 3-3. Guidelines for the Use of the Moon	16
Table 5-1. Comparison of Three Methods for Star Precomputation	33
Table 5-2. A Worst Case Comparison of Star Precomputation	35
Table 5-3. Precomputed Sun, Moon and Venus	37
Table 5-4. Sun's Height (Hc) and Bearing (Zn) on July 14th, 1982	39
Table 5-5. Precomputation of Stars and Planets	42
Table 5-6. More Precomputation of Stars and Planets	45
Table 6-1. Precomputed LOP Intersection Angles Throughout the Day	52
Table 6-2. Sun-Moon Fixes Predicted by the Star Finder	53

Chapter 1.
Introduction

Various devices have been invented to help navigators identify stars and planets and solve other related problems in celestial navigation. The most popular of these was once an official government product called H.O. 2102-D. These are no longer produced by the government, but identical ones are now available from several commercial companies—trade names usually include "2102-D" to identify the product. They cost about $50, and are well worth the investment.

The standard user's instructions that come with these Star Finders, however, are sparse and far too terse for celestial navigation students who are new to the device. We have heard this from nearly every student who has tried to learn the use of this Star Finder on their own. This booklet is intended to correct this shortcoming and extend the discussion of this Star Finder that appears in Bowditch's American Practical Navigator, Vol. I, Article 1340, as well as Chapter 21 of Dutton's (Fifteenth Edition).

It is also in part the fault of the limited user's instructions that this device is not any better known outside of the celestial navigator's circle—a fairly small circle to begin with, that keeps shrinking rapidly as electronic navigation aids increase in convenience and decrease in price. We hope this booklet might add new interest to this valuable tool. The Star Finder is a handy gadget. It could be especially valuable to amateur astronomers or anyone else who spends much time outdoors, looking at the stars, wondering who is who, and where the planets are, and so forth.

To use the Star Finder you will also need a current edition of the *Nautical Almanac*, or its equivalent in the professional or amateur astronomer's fields. You don't, however, have to be an experienced navigator to use the 2102-D Star Finder or this guide. All that is required is some familiarity with the basic terminology of celestial navigation—much of which is reviewed here—and the use of your choice of Almanac. This booklet was prepared for students in the process of learning celestial navigation, but we suspect that even seasoned navigators may find the Star Finder more valuable with this expanded description of its usage.

Again, even though our emphasis throughout is on navigational applications, you don't have to go to sea to benefit from the Star Finder and this book. But if you do plan to go to sea someday, the Star Finder is an excellent way to prepare for your celestial navigation. From your own backyard you can learn the stars you will start out with, and even look ahead to see what stars will be your guides as you approach Tahiti or The Azores in the future.

The Primary Navigational Applications of the Star Finder Are:

(I) To identify stars or planets after taking sextant sights of them—a circumstance that occurs on partly cloudy twilights, when only a few unknown stars are available in isolated patches of clear sky.

(II) To predict (precompute) the heights and bearings of stars and planets before you take sextant sights of them—a standard procedure that makes sights easier and more accurate. Precomputed sights are usually more accurate because, by knowing exactly where to look for the stars, you can take sights earlier in the evening twilight while the sea horizon is still a sharp line. It is the sharpness of the horizon that usually determines the accuracy of sextant sights.

The Star Finder Book

(III) To determine the optimum time of day for simultaneous sights of sun and moon, or, more rarely, sun and Venus—a less common application, but when needed, the Star Finder is a convenient way to solve the problem.

Another application is simply to use the Star Finder to help learn the stars and how they move across the sky. Or, use it to look ahead to see what stars you might use for navigation during a planned voyage in some later season. In short, to use it for whatever you want to know about the positions and apparent motions of celestial bodies.

Once you become familiar with the use of the Star Finder, it becomes, in essence, your own pocket planetarium. You can use it to view the sky during any season, from any latitude, at any time of day or night. As you read through this guide, it will be helpful to have a Star Finder at hand to identify the parts and work through the numerical examples.

Throughout the text there are a few notes on finding directions from the stars. These address various aspects of stargazing for orientation.

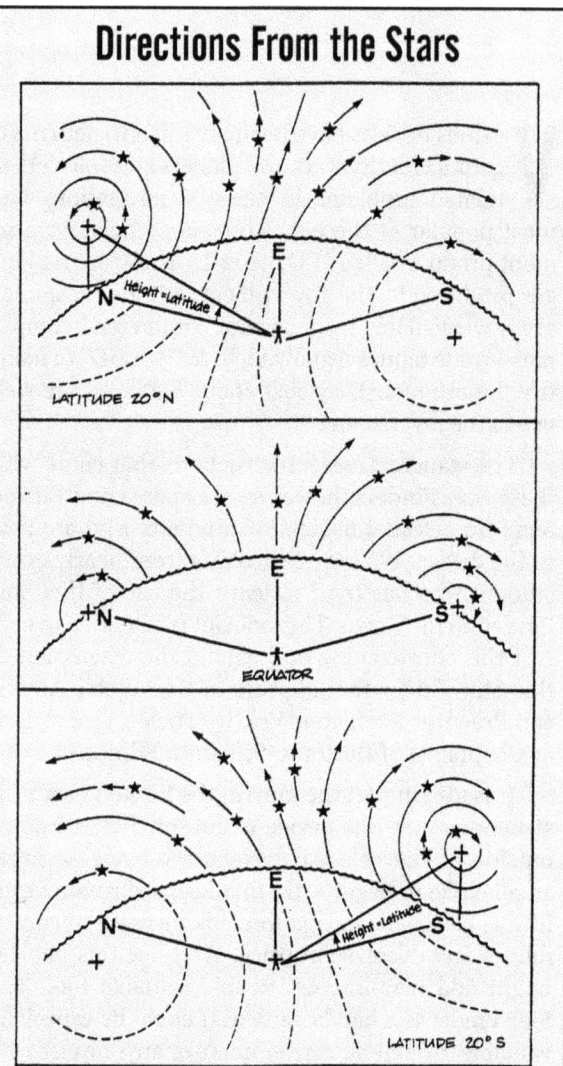

Directions From the Stars

Using pointers to find north when Polaris is obscured. Hold a stick in line with the pointers, mark the distance between them, and then mark five times this distance from the end of the stick. Hold the stick as shown, and you have found north. The weighted string is seldom required.

Adapted from Emergency Navigation, *by David Burch, (McGraw-Hill, 1986, 2008). See related discussion in Chapter 7 of this book*

Directions From the Stars

THE APPARENT PATHS OF STARS (SHOWN SCHEMATICALLY) AROUND THE ELEVATED CELESTIAL POLES AS SEEN FROM NORTHERN AND SOUTHERN LATITUDES. Each star circles the pole once every 24 hours. Stars that never rise or set as they circle are called circumpolar stars; they remain in view all night long, every night of the year.

Adapted from Emergency Navigation, *by David Burch, (McGraw-Hill, 1986, 2008). See related discussion in Chapter 7 of this book*

Chapter 2.
Description of the 2102-D Star Finder

The device consists of three parts: (1) A rigid white plastic disk with stars printed on it, called the "star base" or "white disk," (2) a set of 9 transparent templates with blue printing, which we call the "blue templates," and (3) a special template with red printing. These parts, and a limited set of instructions, come in a round black plastic case about a foot in diameter.

The stars printed on the white disk are the 57 "navigational stars" listed on the daily pages of the *Nautical Almanac* (see appendix). One side of the white disk shows the North Pole of the sky at its center; the other side shows the South Pole. The same stars are shown on each side, but from a different perspective. The brightness of a star on the white disk is indicated by its symbol, as shown in Table 2-1.

Special terms like "navigational star," "magnitude" of a star, "declination" of a star, and a few more used in this section, are explained later in Sec. 3.1 on terminology.

The type of star map used for the stars on the white disk is similar to the round star maps at the back of the *Nautical Almanac*, although the layout is slightly different, as shown in Figs. 2-1 and 2-2. Looking at the North side of the white disk, stars near the rim of the disk have high southern declinations; those about halfway toward the center have declinations near 0°; and those near the center of the disk have high northern declinations.

The scale along the rim of the white disk represents different things depending on the application.

Table 2-1. Star Symbols on the White Disk

Symbol	Name	Description
⊙	Capella	Northern sky examples: Capella and Vega Southern sky examples: Sirius and Canopus These are "magnitude one" stars, the 20 brightest stars in all the sky. All 20 are navigational stars.
●	Hamal	Northern sky examples: Dubhe and Alkaid Southern sky examples: Nunki and Hamal These are "magnitude two" stars, which are stars as bright as the Big Dipper stars. There are some 75 of the se in all the sky. About 30 are navigational stars.
•	Enif	Northern sky examples: Schedar and Enif Southern sky examples: Gienah and Sabik These are "magnitude three" stars, which are stars not quite as bright as Big Dipper stars, but brighter than the stars on the handle of the Little Dipper. There are some 200 magnitude three stars in all the sky. Only 7 of the brightest of these are navigational stars.

The Star Finder Book

Using earth coordinates, the rim scale represents the Local Hour Angle of Aries (Sec. 3.1); using sky coordinates, the rim scale represents Right Ascension, which is the same as 360° minus Sidereal Hour Angle. But don't worry for now about the technical meanings of this scale and how they are related, because once we have the Star Finder set up for a given location and date, this rim scale will simply become a 24-hour clock face, with each 15° corresponding to 1 hour of time as shown in Figure 2-3.

The positions of the sun, moon, and planets are not shown on the white disk because their positions among the stars change daily. The white disk, however, can be written on with pencil (and easily erased), so we can add these positions to the disk once we look up their sky positions (declinations and Sidereal Hour Angles) in the *Nautical Almanac*. We can also add other stars if we choose, but this is not a common practice. Most important to note is this: we can identify all stars with this device, even those that are not printed on the white disk.

To use the Star Finder we place the appropriate blue template (the choice depends on our latitude) on the pin at the center of the white disk. The blue template is then rotated to set it for the proper time, and then the heights and directions of the stars can be read directly from scales on the blue template. In rare cases, this simple operation of placing the template on the centerpin of the white disk is not as trivial as it should be—especially on some new units, or newly used templates as you sail into new latitudes. On some new units the pin fit is very tight. You might end up commenting on the designer as you try to get the thing on there. But the fit must be snug if the readings are to be accurate. If the fit happens to be tight at first, don't drill the holes bigger, they will loosen with usage.

One approach to getting a tight template onto the pin is to lay the white disk on a hard flat surface, and gently press the template over the pin, pressing on opposite sides near the pin. But don't press too hard because the pin can pop out of the white disk. If you lose the pin, you end

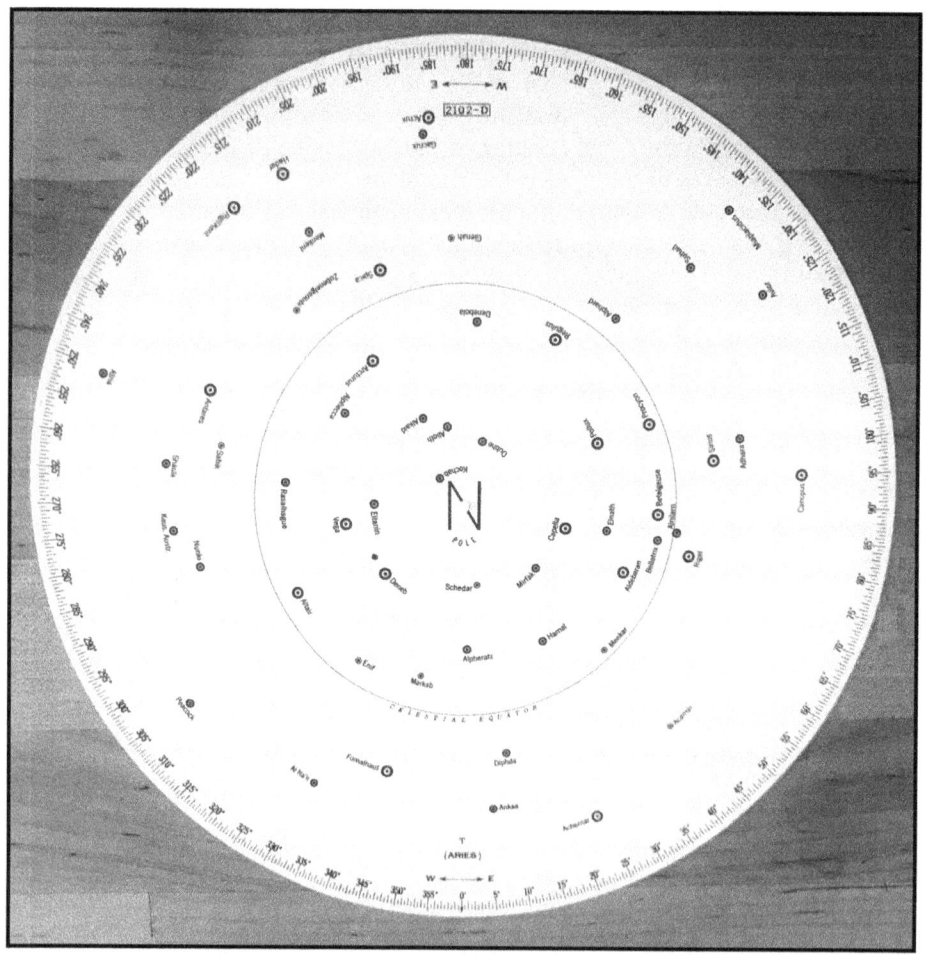

Figure 2-1. *North side of the white disk of the 2102-D Star Finder*

Chapter 2: Description of the 2102-D

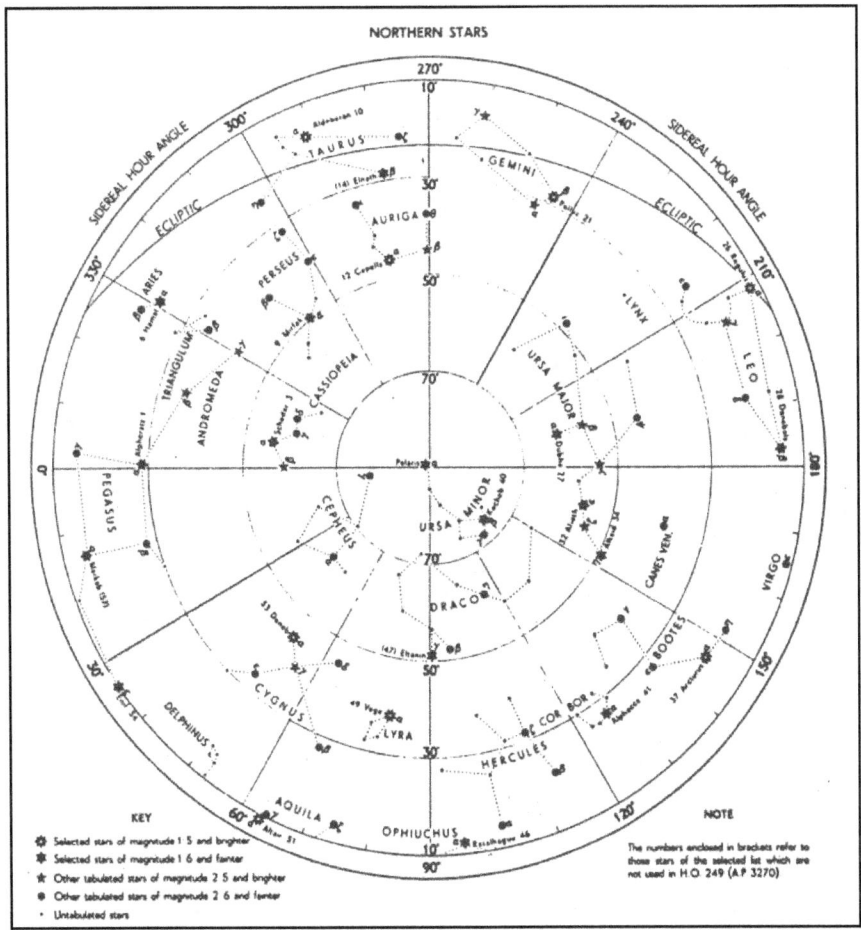

Figure 2-2. *Round star map of the northern stars from the* Nautical Almanac. *(Larger Views in the Appendix.)*

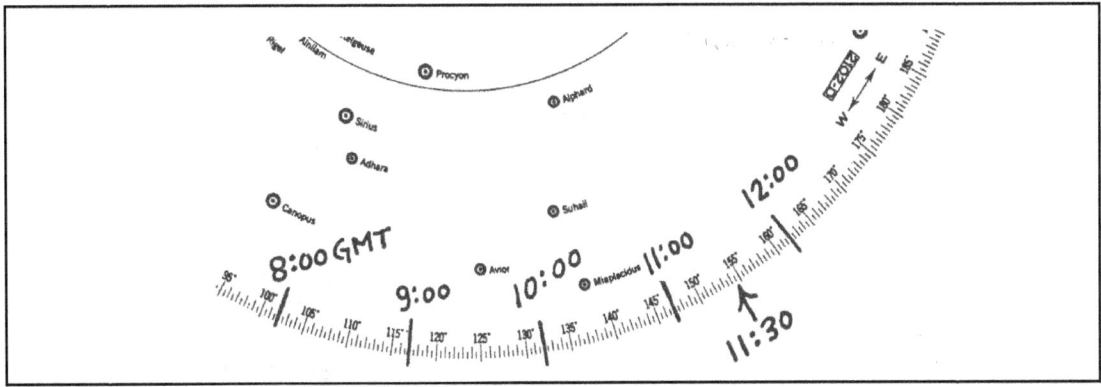

Figure 2-3. *Time scale on the white disk. After one point on the rim scale is assigned a specific time, the remaining time scale follows at 15° per hour. You can use any time scale: watch time, zone time or UTC.*

5

up using a wooden match, or some such thing, in its place—which works, but is more trouble.

The 9 blue templates are labeled according to the latitude of the observer. There is one template for every 10° of latitude, beginning at latitude 5°. For most applications at sea we use the blue template that corresponds most closely to our dead reckoning (DR) latitude—for practice on land, use the one closest to your known latitude

There is a North and South side to each blue template. The north side of the 45° template, for example, is the side labeled "Latitude 45° N." You can see the label for the south side through the transparent template, but it will appear backwards. For all applications in the Northern Hemisphere use the north side of the white disk with the north side of the blue template facing upward so you can read its label properly. When viewing stars from the Southern Hemisphere use the south side of the blue template on the south side of the white disk.

Each blue template contains two overlapping sets of curves. One set represents the angular heights of the stars; the other set gives the true bearings of the stars. In the terminology of celestial navigation, the height of a star would be its calculated altitude (Hc), and its true bearing would be its azimuth (Zn). The same printed scales apply to all celestial bodies: sun, moon, stars, and planets. The azimuth scale is the one that goes from 0° to 360° in 5° intervals around the diagram. The altitude scale goes from 0° (on the horizon) to 90° (overhead) in 5° intervals across the diagram. The proper numerical scales to use are the ones that can be read properly. Numerical scales for the other side of the template can be seen through

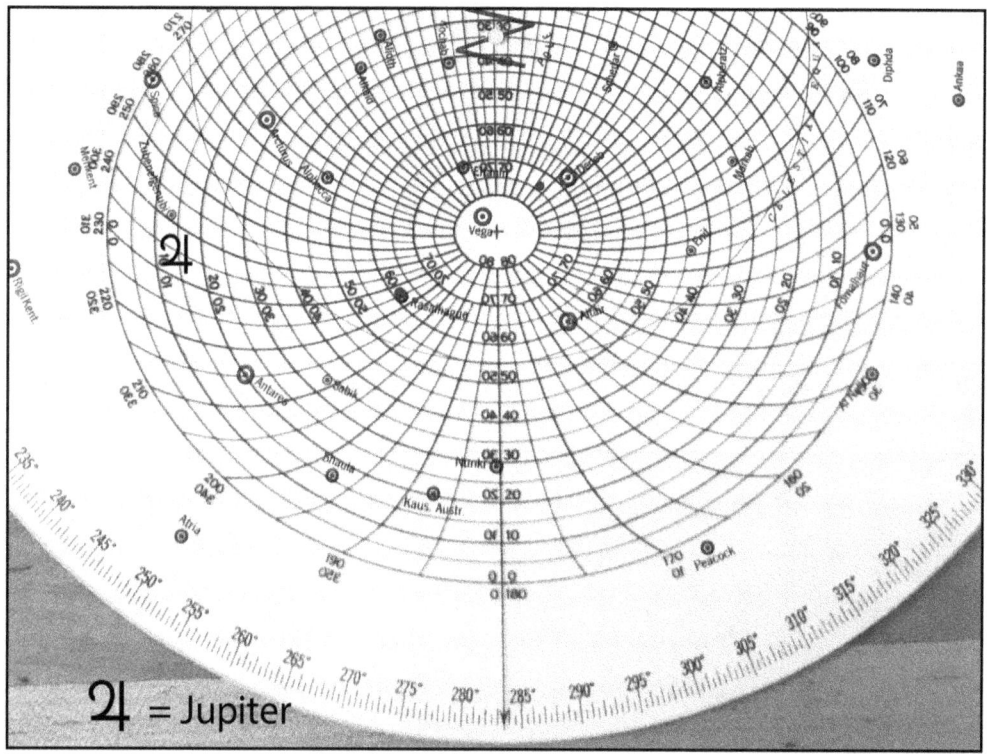

Figure 2-4. *The blue template for 35° N in place on the north side of the white disk. The blue-template arrow is set to Local Hour Angle Aries = 283.5°, which, in this example, corresponds to a UTC of about 15:30. The star Vega is near overhead, bearing to the NW; Antares bears about 215° at a height of 20°. Arcturus bears 276° at a height of about 27°. Spica is just below the horizon to the west, and Fomalhaut is rising at about 130°. Altair is about 62° high, bearing about 151°. The planet Jupiter (plotted by hand) is about 15° high, bearing 235°. All Star Finder bearings are True bearings.*

Chapter 2: Description of the 2102-D

it, but they appear backwards. Figure 2-4 shows a blue template in place on the white disk and lists a few heights and bearings that can be read from these scales.

The red template is a special one used to plot sun, moon, and planet positions onto the white disk. It is also used for identifying sighted stars that do not appear on the white disk. The concentric circles on the red template correspond to lines of equal declination on the white disk. Circles on the center part of the red template are solid lines; the outer ones are dashed lines. We will later use this difference to distinguish between north and south declinations when identifying an object that is not printed on the white disk. A declination scale is printed along a radial line. The slot in the red template is used for plotting sun, moon, and planet positions on the white disk.

The scale along the rim of the red template can be ignored; it has no significant practical use. There are also north and south sides labeled on the red template, but since the rim scale is not needed for any practical application, we can use either side of the red template for all applications. The south side of the red template, however, will prove the more convenient for all applications because of the way the declination scale is printed.

2.1 The Original Navy Star Finder

The 2102-D Star Finder started off as a somewhat different looking device, patented in 1921 by G. T. Rude and sold to the Navy where it was distributed as HO 2102-A. Figure 2-5 shows the parts. It was notably larger at 17" across when closed, presented in a foldout format with paper base plates and two Lat templates per sheet on clear plastic.

Notable differences beyond size were its inclusion of time and date scales (with leap year corrections) so an almanac was not required, and tiny inserted corrections printed in red across the white baseplate page such as (–5/4). The top figure is the correction in altitude per degree of Lat away from the template label, and the bottom figure is the correction to azimuth in tenths of a degree per degree of Lat away from the template value. We address these corrections using 2102-D in Section 7.2. Like the 2102-D, there is a template for every 10° of Lat.

2.2 British Star Finder, NP 323

Although it looks different, the British Star Finder NP 323 is essentially the same as the 2102-D. Explanations in this book apply to it as well. The latitude templates are orange and only printed on one side. Both star finders give templates for every 10° of latitude, but NP 323 starts at Lat 0°, not 5°. The white disk of 2102-D is replaced with a white paper card in NP 323 and there is no pin in the center, which makes its alignment and rotation slightly more difficult. The sizes of the scales are the same, so they have the same accuracy. There is no equivalent to the 2102-D's red template. In short, the two are the same for practical purposes, although the 2102-D is slightly easier to use and more durable.

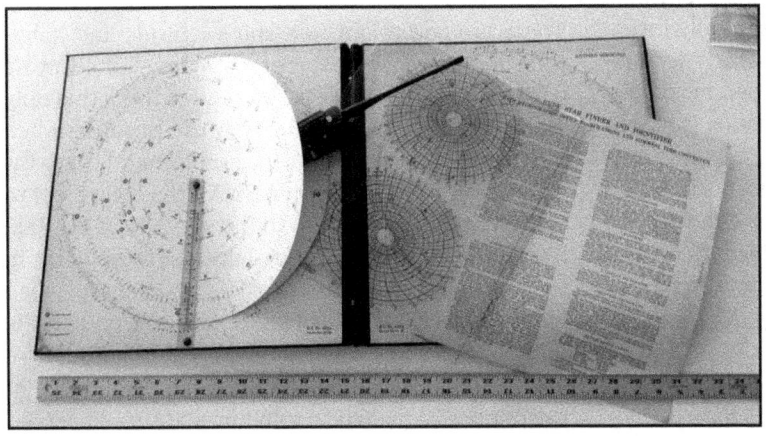

Figure 2-5. *HO 2102-A, "Rude Star Finder and Identifier, with Hydrographic Office Modifications and Sidereal Time Converter," December, 1921 edition. A detailed description of this historic device can be seen at davidburchnavigation.blogspot.com/2019/08/2102-A.html.*

Chapter 3. Background

3.1 Terminology and Celestial Motions

The 57 stars printed on the white disk are the so-called "navigational stars." These are, by definition, the stars listed on each of the daily pages of the *Nautical Almanac*—the same 57 stars appear on each page. These include the 20 brightest stars, but the remaining 37 are not the next brightest ones. The remaining stars were chosen because of their position, as well as their brightness, to provide a uniform distribution of stars across the skies for navigational use. You could, if necessary, navigate by any one of several hundred stars visible on a clear night, but it is more convenient to use the navigational stars whenever possible. The *Nautical Almanac* lists (at the back of the book) the necessary star data for another 100 or so stars, but these are not officially called navigational stars.

A "northern star" is a star with a north declination, meaning it circles the earth above the Northern Hemisphere at a latitude equal to its declination. This is the definition of declination; it is the latitude over which the star circles. Each star circles the earth over its own unique latitude. And this latitude, its declination, remains constant, essentially forever. If your latitude equals a particular star's declination, then that star must pass directly overhead once a day, all year long. During some seasons it passes during daylight hours so you won't see it, and in other seasons it is visible at night but doesn't actually go over head at night, but it goes by once a day nevertheless. A "southern star" has south declination and circles the earth over the Southern Hemisphere.

From either hemisphere, we can see both northern and southern stars. Looking low in the sky from any latitude, north or south, we see northern stars to the north, and southern stars to the south, and both northern and southern stars when looking east or west. The stars we see overhead depend on our latitude. In northern latitudes, only northern stars pass overhead; in southern latitudes, only southern stars pass overhead. Any star that passes precisely overhead (crosses our zenith) must have a declination equal to our latitude. Now a fact that is not at all obvious: From any latitude, north or south, southern stars always rise south of due east and set south of due west. Likewise, northern stars rise north of east and set north of west. This is the reason we see both northern and southern stars when looking east or west from any latitude. Without the use of a planetarium it is somewhat difficult to picture how this comes about, but Figure 3-1 might help clarify this. Soon we will be able to see this directly from the Star Finder. This can be a handy thing to know when it comes to star ID.

With the use of the Star Finder, every object in the sky can be precisely located once we know its declination (Dec) and sidereal hour angle (SHA). These are equivalent to a star's latitude and longitude on a star map or star globe. When first learning celestial navigation, we might tend to think that only stars have an SHA, since we only use this SHA when navigating with stars. But every object and every vacant point in the sky has an SHA. It is equivalent to the longitude of the point in the sky, just as the declination is equivalent to the latitude of the point in the sky. Every point in the sky is uniquely specified by its SHA and Dec, just as every point on earth is uniquely specified by its latitude and longitude. We will use this concept when discussing the location of the moon and planets among the stars.

The line in the sky with Dec = 0° is called the celestial equator; it circles the earth directly above the earth's equator. The celestial meridian with SHA = 0° is called "Aries," or, sometimes, the "First Point of Aries." It is equivalent to the Greenwich meridian on earth. As the earth rotates daily beneath the stars, the Aries meridian, running from the north pole of the sky to the south pole of the sky, slides westward around the earth. It lies directly above the Greenwich meridian once every 24 hours. Earth and sky coordinates are illustrated in Figure 3-2.

Unfortunately, there is a slight difference in the terminology used by astronomers and navigators. Both use Dec in the same way to specify the latitude of a star, but astronomers do not use SHA for longitude. They specify celestial longitude by "right ascension." SHA is measured in degrees west of Aries, whereas right ascension is measured in time units (1 hour for each 15°) east from Aries. We will stick with the navigator's usage of SHA, but astronomers should have no difficulty in adjusting for this distinction throughout this book. To convert from right ascension to SHA:

SHA = 15° x (24 hr – right ascension).

Astronomers should find the Star Finder easy to use, since the rim scale on the white disk is simply right ascension expressed in degrees.

Positions on the white disk can be pictured in terms of the (invisible) Dec-SHA grid. The star Deneb, for example, with Dec = N 45° and SHA = 50° appears on the "north side" of white disk about one inch from the center, along the radial line from the center to the 310° position on rim scale (360 – SHA). On the "south side" this star appears at the same location on the rim, but with the south pole at the center, it appears about one inch from the rim. To help understand the layout of this star map, compare the Deneb

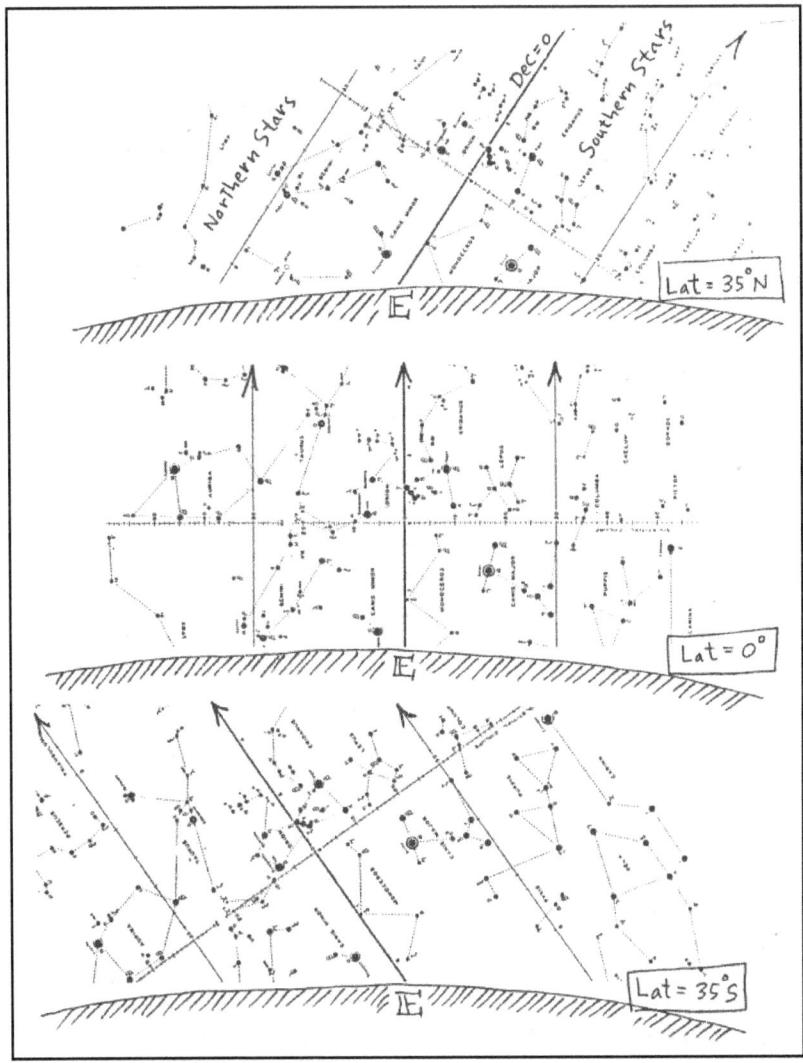

Figure 3-1. *Rising and setting stars. Viewed from any latitude, north or south, southern stars (or any object in the sky with a south declination) always rise south of due east and set south of due west. Likewise northern stars rise north of east and set north of west. Celestial paths across the sky are always symmetric about the meridian. If a star, or the sun, rises 15° north of east, it will set 15° north of west.*

Table 3-1. Brightness vs. Magnitude	
Magnitude Difference	Brightness Difference
0.0	1.0
0.2	1.2
0.4	1.4
0.6	1.7
0.8	2.1
1.0	2.5
1.2	3.0
1.4	3.6
1.6	4.4
1.8	5.2
2.0	6.3
2.2	7.6
2.4	9.1
2.6	11
2.8	13
3.0	16
3.2	19
3.4	23
3.6	28
3.8	33
4.0	40
4.5	63
5.0	100
5.5	158
6.0	251
6.5	398

position on the white disk (Figure 2-1) to its position on the round star map from the back of the *Nautical Almanac* (Figure 2-2).

The sun and stars are essentially fixed in space. But since the earth circles the sun, the stars behind the sun, as viewed from the earth, change continuously throughout the year. The positions of the moon and planets among the stars also change from day to day because of their orbital motions. As a result of these complicated motions, the sun, moon, and planets all appear from earth to move through the stars from day to day. This means their positions on the white disk change daily. To use the Star Finder for these bodies we must first mark their positions on the white disk and then occasionally replot them to keep their positions accurate.

Of the five planets visible to the naked eye (Mercury, Venus, Mars, Jupiter, and Saturn), Venus and Jupiter are the two we care about the most for navigation. The daily motions of these two, however, are quite different.

Jupiter moves much slower than Venus. It takes Jupiter about a year to move the length of one constellation. Its general direction is eastward through the stars, but its path across the white disk is not a simple straight line. To know precisely where it is you must check its position once a week or so. The season and times we see Jupiter depend on the season of its neighboring stars.

The rate that Venus moves through the stars varies throughout the year. It can move as fast as one constellation per month, so it must be checked every few days or so. Venus also moves basically eastward through the stars, but its apparent motion relative to the sun—which is important because this determines when we see it—is more complex. Venus moves back and forth, from one side of the sun to the other, about twice a year. When Venus is east of the sun (trailing it), it follows the sun over the western horizon, and we see it in the evening just after sunset. When west of the sun, Venus rises before the sun and we see it as a morning "star." In this case it sets before the sun so we can't see it in the evening. The Planet Notes at the front of the *Nautical Almanac* give a brief description of the location and visibility of the planets for the year of issue.

The sun moves through the stars at the rate of 1° per day. It follows—year after year—a path called the "ecliptic," which is drawn on the star charts at the back of the *Nautical Almanac*. The moon and planets also follow very nearly along this path. The band of stars along the ecliptic make up the 12 zodiac constellations. Consequently, the sun, moon, and planets are always found within some zodiac constellation.

The moon is the most evasive of all celestial bodies. It moves eastward through the stars at a rate of about 12° per day, or 30' per hour. In other words, if the moon is directly above the star Aldebaran on one night at, say, 10 p.m., the next night at 10 p.m. it will appear, with a slightly altered phase, about a hand width to the east of Aldebaran; the next night two hand widths to the east, and so on, as shown in Figure 3-3. For accurate moon prediction with the Star Finder, this means you must plot it for very nearly the right time you plan to look for it. An alternative is to plot its position for every 12th hour or so, and connect the dots. You can then estimate its position from the time of day relative to the times at the dot positions.

One way to get a quick perspective on where each of these bodies is relative to the stars, or relative to the sun, is to plot the sun, moon, and planets on the rectangular star maps at the back of the *Nautical Almanac*. These rectangular maps of the ecliptic re-

Chapter 3: Background

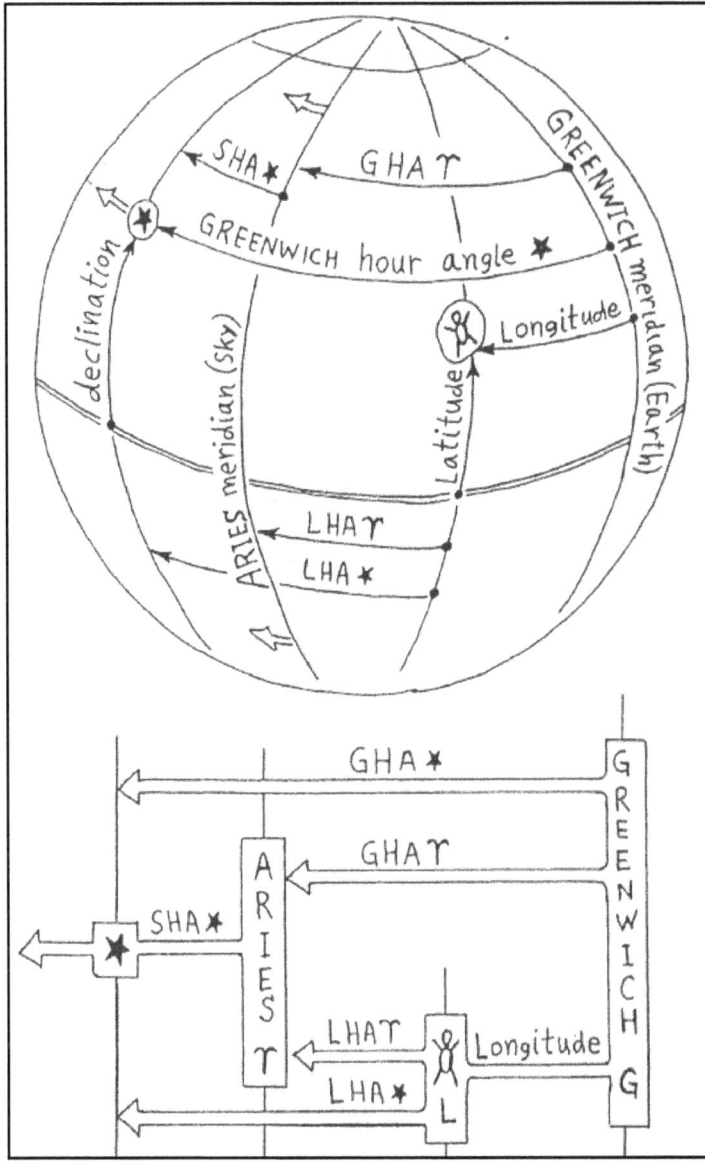

Figure 3-2. *Earth and sky coordinate systems. The sky coordinates are shown projected down onto the earth. Greenwich Hour Angles are measured relative to the Greenwich meridian (G); Local Hour Angles are relative to the observer's meridian (L). Our longitude is how far we are west of the earth's reference meridian (Greenwich); Sidereal Hour Angle is how far the star is west of the sky's reference meridian (Aries).*

gion of the sky are Mercator projections that provide a realistic view of the relative star positions. The sun, moon, and planets can always be located on these maps as shown in Figure 3-4.

These maps will not give heights and bearings, as the Star Finder does, but nevertheless they are convenient for locating the planets relative to the stars, and the moon relative to the sun. Remember that these maps "slide across the sky" at a rate of 15° of SHA each hour. The maps are labeled directly with Dec and SHA, which simplifies the plotting.

3.2 Brightness and Color of Stars and Planets

The brightness of a star is often a valuable aid to its identification. The brightness of a star or planet is determined by its "magnitude." Star magnitudes are given in the extra star list at the back of the Almanac. They are not listed on the daily pages, but the card insert to the Almanac has a convenient list. This card is reproduced on a page at the back of the Almanac, shown here in Figure 3-5. Planet magnitudes are listed at the heads of the planet columns on each daily page because their brightness changes slowly throughout the year. The same magnitude scale applies to stars and planets.

Unfortunately, there is not a simple correspondence between the numerical magnitude of a star and the visual brightness that we perceive. Each magnitude difference of 1.0 implies a brightness difference of 2.5. The magnitude scale is logarithmic, which means we need special tables, such as Table 3-1, to figure the actual brightness difference between two stars, or between a star and planet. And to complicate things even further, the scale is inverted; the lower the magnitude number, the brighter the star.

The faintest stars we might navigate by would have a magnitude of about 3.0 although it would be rare to use such a faint star. A typical bright star has magnitude 1.0, which we could say is 2.0 "mag-

The Star Finder Book

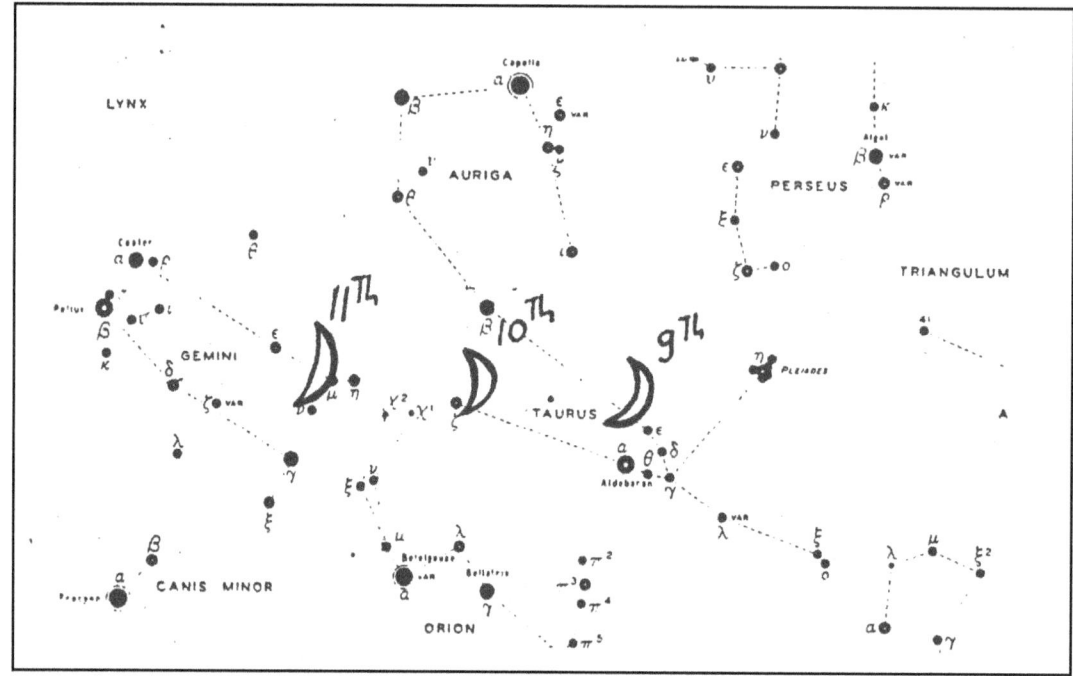

Figure 3-3. *Nightly motion of the moon relative to the stars.*

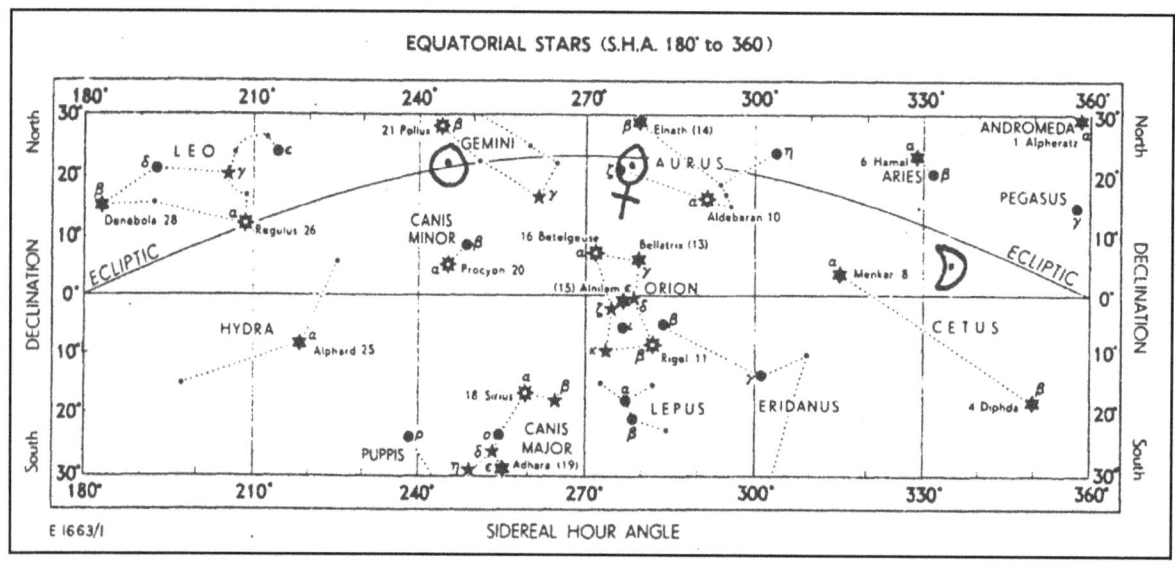

Figure 3-4. *Rectangular star maps from the* Nautical Almanac. *These provide a quick means of locating the relative positions of sun, moon, and planets. The plotted positions are for July 14th, 1982. (Larger views in the Appendix.)*

Chapter 3: Background

	1982 DECEMBER 21, 22, 23 (TUES., WED., THURS.)									
	ARIES	VENUS −3.4		MARS +1.3		JUPITER −1.3		SATURN +0.9	STARS	
G.M.T.	G.H.A.	G.H.A.	Dec.	G.H.A.	Dec.	G.H.A.	Dec.	G.H.A. Dec.	Name	S.H.A. Dec.
21 00	89 14.2	168 01.1	S23 59.0	138 05.7	S19 21.8	212 18.4	S19 07.6	238 31.9 S 9 58.4	Acamar	315 35.7 S40 22.6
01	104 16.7	183 00.1	58.9	153 06.1	21.3	227 20.3	07.8	253 34.2 58.5	Achernar	335 43.7 S57 19.8
02	119 19.1	197 59.2	58.7	168 06.6	20.7	242 22.2	07.9	268 36.4 58.6	Acrux	173 35.6 S62 59.9
03	134 21.6	212 58.2	58.5	183 07.1	20.2	257 24.2	08.0	283 38.7 58.6	Adhara	255 30.5 S28 56.8
04	149 24.1	227 57.2	58.4	198 07.5	19.7	272 26.1	08.1	298 40.9 58.7	Aldebaran	291 15.8 N16 28.5
05	164 26.5	242 56.3	58.2	213 08.0	19.2	287 28.0	08.2	313 43.2 58.8		

INDEX TO SELECTED STARS, 1982

Name	No.	Mag.	S.H.A.	Dec.	No.	Name	Mag.	S.H.A.	Dec.
Acamar	7	3.1	316	S. 40	1	Alpheratz	2.2	358	N. 29
Achernar	5	0.6	336	S. 57	2	Ankaa	2.4	354	S. 42
Acrux	30	1.1	174	S. 63	3	Schedar	2.5	350	N. 56
Adhara	19	1.6	256	S. 29	4	Diphda	2.2	349	S. 18
Aldebaran	10	1.1	291	N. 16	5	Achernar	0.6	336	S. 57
Alioth	32	1.7	167	N. 56	6	Hamal	2.2	328	N. 23
Alkaid	34	1.9	153	N. 49	7	Acamar	3.1	316	S. 40
Al Na'ir	55	2.2	28	S. 47	8	Menkar	2.8	315	N. 4
Alnilam	15	1.8	276	S. 1	9	Mirfak	1.9	309	N. 50
Alphard	25	2.2	218	S. 9	10	Aldebaran	1.1	291	N. 16

Figure 3-5. *Star and planet magnitudes listed in the* Nautical Almanac *(page xxxiii).*

nitudes" brighter than the faint magnitude 3 star. But the actual brightness difference between the two would not be a factor of 2.0; the bright one would appear about 6 times brighter than the faint one.

The magnitude scale can also go negative for very bright objects. Venus, for example, at magnitude −4.0 would be 5.5 magnitudes brighter than a star with magnitude 1.5. Only two stars, the southern stars Sirius and Canopus, are bright enough to have negative magnitudes. Venus and Jupiter are always negative, meaning always very bright, but Mars and Saturn are only rarely negative.

Note: the sign of the magnitude difference is not important; the object with the smaller magnitude is always the brighter object. Remember -1 is "smaller" than +1; and −3 is "smaller" than −2, and so forth. Objects with the same magnitude are equally bright. In Table 3-1 this is indicated by showing that a 0 magnitude difference means an object is 1.0 times brighter than another object with the same magnitude.

For star identification it is not necessary to be very technical about brightness and magnitudes. It is sufficient to classify stars in 3 rough categories: "magnitude one" stars, the 20 or so brightest ones (pick a favorite and use it as your standard); "magnitude two" stars, the stars about as bright as the Big Dipper stars; and "magnitude three" stars like those on the handle of the Little Dipper. The vast majority of celestial navigation is done with magnitude one stars, and magnitude three stars are hardly ever used.

Table 3-2. Distinctly Reddish[a] Navigational Stars

Mag. 1[b]		Mag. 2		Mag. 3	
Betelgeuse	1.87[c]	Suhail	1.65	Enif	1.55
Antares	1.84	Gacrux	1.55	Schedar	1.18
Aldebaran	1.52	Eltanin	1.52	Markab	1.02
Arcturus	1.23	Kochab	1.47		
Pollux	1.02	Alphard	1.44		
		Atria	1.43		
		Hamal	1.15		
		Ankaa	1.08		
		Dubhe	1.06		

(a) Including yellowish and orangish. Listed in descending order of "redness."
(b) Grouped according to the approximate magnitude scale discussed in the text.
(c) These numbers are an indirect measure of their "redness"—the bigger the number the redder the star, but the differences are slight. Two stars with the same number, however, should appear about the same color.

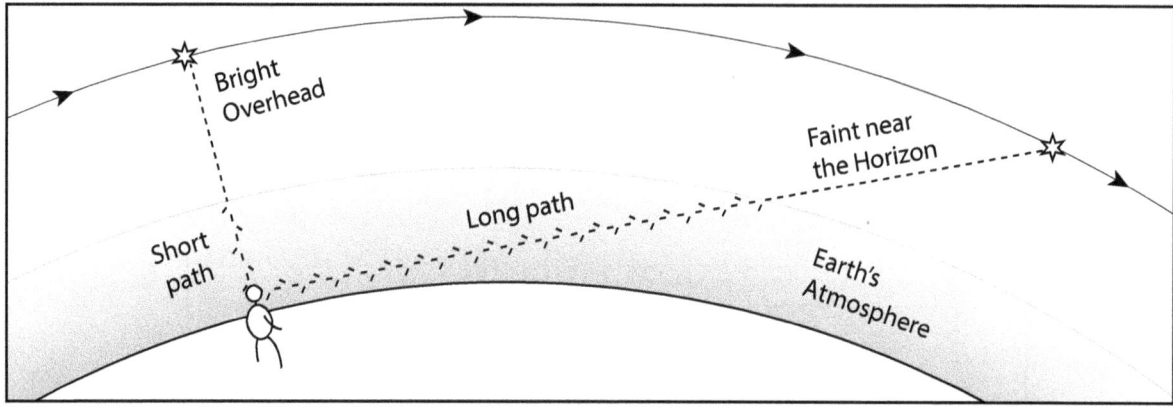

Figure 3-6. *How star brightness changes with the height of the star. All stars fade as they descend toward the horizon because more of their light is lost to scattering.*

A side tip on star ID: It is rare to see stars low on the horizon—say, within a hand width or so—because there we view them through the thickest layer of the earth's atmosphere, where much of their light intensity is lost to scattering. Even the brightest stars fade as they descend toward the horizon, as shown in Figure 3-6. Consequently, if you see an isolated star low on the horizon, you can bet it is a bright one, even if it appears faint. Since bright stars are well known stars, this observation alone often identifies the star for you.

Or, an isolated low "star" could be Venus or Jupiter. But this confusion is unlikely since navigators tend to keep pretty close track of where these guys are, and even low on the horizon they remain notably bright. On a clear night, a low, bright Venus can startle a weary helmsman who sees it for the first time.

For more sophisticated star ID, it helps to know that several stars are distinctly reddish. These are in a class of stars called the "Red Giants," and knowing these can be a valuable aid to their identification. Table 3-2 is a list of stars that most observers can routinely recognize as distinctly reddish—or more precisely, will later recognize as red once they have been pointed out as red and compared to neighboring white stars.

Example 3-1
Figuring the Brightness of Stars and Planets

(1) Arcturus has magnitude 0.2 and Dubhe has magnitude 2.0. The magnitude difference is 2.0 − 0.2 = 1.8, and from Table 3-1, Arcturus is 5.2 times brighter than Dubhe.

(2) Sirius has magnitude -1.6 and Antares has magnitude 1.2. The magnitude difference is 1.2 − (−1.6) = 1.2 + 1.6 = 2.8, and from Table 3-1, Sirius is 13 times brighter than Antares.

(3) Jupiter, on some date, has magnitude −2.1 and Canopus has magnitude -0.9. The magnitude difference is −2.1 − (−0.9) = −2.1 + 0.9 = −1.2, and from Table 3-1, Jupiter is 3 times brighter than Canopus.

(4) Venus can be as bright as magnitude −4.3 and the North Star, Polaris, has magnitude +2.1. The magnitude difference is 2.1 − (−4.3) = 6.4. From table 3-1 we can estimate that Venus is roughly 400 times brighter than Polaris.

3.3 Tips on Planet Identification

The planet Mercury can be seen with the naked eye, and it can even be quite bright. But it is only rarely visible, low on the horizon, just before sunrise or just after sunset, very near the sun. Since it is rare to be seen and always very low on the horizon it is not used for navigation. Its Dec and SHA are not listed in the *Nautical Almanac*.

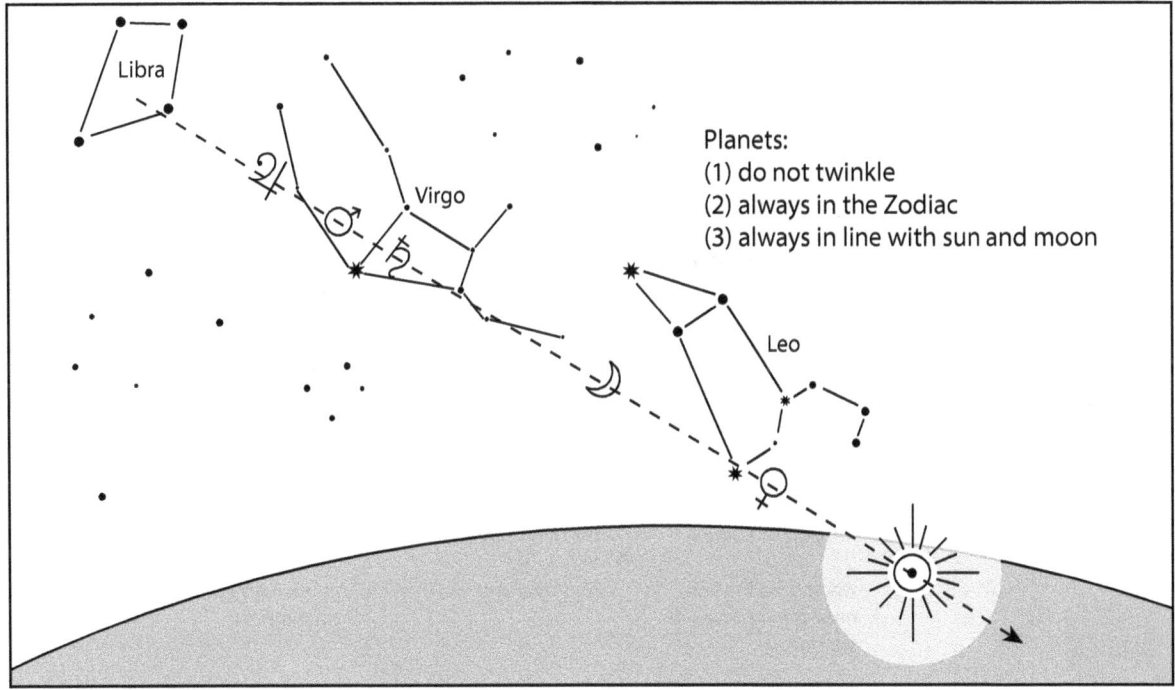

Figure 3-7. *Helpful hints for planet identification.*

Venus and Jupiter always stand out nicely among the stars. When either of these two are visible, they are always much brighter than any stars around them. Mars and Saturn, on the other hand, appear similar to bright or medium-bright stars. The main function of Mars and Saturn is to confuse the navigator by appearing as stars where no stars should be. Mars can sometimes appear reddish, and most planets will appear as tiny disks, rather than points, when viewed through 10x magnification binoculars.

Another identifying characteristic of planets is their lack of twinkle. Stars twinkle, planets do not. The reason can be traced to the size of the light source—stars are point sources of light; planets are disk sources. A patch of warm air can momentarily refract all of the star light out of our eye, causing it to twinkle; but such transient refraction cannot remove all of the light from a planet since it comes from slightly different angles depending on its origin on the disk.

The relative location of the planets can also sometimes confirm or assist in their identification. The sun, moon, and planets always lie along the same arc across the sky. On those occasions when 3 or more of these objects are visible (say, moon and two planets), this alignment can sometimes aid in their identification. Again, planets always lie within a zodiac constellation. These tips are summarized in Figure 3-7.

3.4 Use of the Moon and Planets

The accuracy of twilight sextant sights is typically determined by the sharpness of the horizon. If the horizon is a sharp line we get good sights because we know precisely where to align the stars in the sextant. When the horizon is obscure, with the sky blending almost imperceptibly into the sea, we must resort to our best judgment when aligning the star with the horizon—and this will vary from person to person, and from one star to another for the same person looking in different directions.

Evening sights start with a good horizon and end with a poor one. The reverse is true for morning twilight sights: we start in the dark with a weak horizon which then slowly sharpens as daylight approaches.

Table 3-3. Guidelines for the Use of the Moon[a]

Age (days)	Phase and Location	Sight Time	Special Value
1-3	waxing crescent, setting near western horizon at sunset	just after sunset, before evening stars appear	combine with evening star sights
6-8	waxing half moon, near the meridian sunset	mid afternoon	combine with sun at for day-time fix
12-14	waxing full moon, rising near eastern horizon at sunset	just after sunset, before evening stars appear	combine with evening star sights
15-17	waning full moon, setting near western horizon at sunrise	just before sunrise, after morning stars fade	combine with morning star sights
21-23	waning half moon, near the meridian at sunrise	mid morning	combine with sun for daytime fix
26-28	waning crescent, rising near eastern horizon at sunrise	just before sunrise, after morning stars fade	combine with morning star sights

(a) *The range of ages might be off a day or so in some circumstances. The predictions are least dependable when either sun or moon pass overhead during the days in question. The best use of the moon depends on your latitude and the declination of the moon. The moon's age is listed on the daily pages of the* Nautical Almanac.

Any sights we can take during the brighter part of twilight, morning or evening, will improve the accuracy of our fixes. The moon, Venus, and Jupiter are the three bodies that can be seen during the brightest part of twilight when the stars are not visible.

Since we need a triangle of three well positioned bodies for a strong (good) fix, the standard procedure is to take, whenever possible, whatever is available from Venus, Jupiter, or possibly the moon and then pick the best stars available to complete the triangle. Most navigators would probably agree with this use of Venus or Jupiter, but all might not extend this philosophy to the moon, since sights of a bright moon might not be as accurate as those of a well-positioned crescent moon or a planet. But regardless of these details in philosophy, the procedure is the same. It is illustrated with numerical examples in Chapter 5.

Sometimes none of these bright guys will be available, or the moon will be there but not in a usable phase or location. In that case we go by stars alone, but even then it is always best to figure ahead of time the best three stars to use, and precompute their heights and bearings. In the evening, this procedure still gives you a good head start on the brighter stars before the horizon begins to fade; in the morning, it lets you postpone brighter stars to the end of twilight when the horizon has improved. The procedure is explained in Chapter 5.

In summary, if Venus or Jupiter is available it is good practice to use them. They are bright enough to spot by just looking around as you start your cruise. To find out where they will be and when they might be useful for a later cruise, check the *Nautical Almanac* Planet Notes. As a general rule, Mars and Saturn offer no special aid to your sights, and they would only be chosen if they happen to be bright and make up the best triangle with the available bright stars. Procedures and criteria for choosing stars or star-planet combinations are discussed in Chapter 5.

The use and usefulness of the moon cannot be specified precisely because it is difficult to make reliable generalizations about where the moon is. It simply moves around too much in the sky, and as a result its location in your sky depends on when and where you are. But we can provide guidelines which we have found are quite valuable for getting started at more precise predictions. These guidelines are presented in Table 3-3. They are given according to the age of the moon which is listed for each day on the daily pages of the *Nautical Almanac*.

Besides this table of guidelines, you also have the moonrise and moonset times given in the *Nautical Almanac*. When considering moon sights, first check Table 3-3 then double check the moon's rising and setting times to be sure the moon is above the horizon at the time you plan the sights. The meridian passage time of the moon is also listed. Use the time of Upper Transit to tell when the moon will cross your meridian—when it will be at its maximum height in the sky, bearing either due north or due south depending on your latitude and the declination of the moon. All times listed are Local Mean Times (explained in the next section), but for judging roughly where the moon is relative to the sun, you can consider 1200 LMT to be midday. If the moon's meridian passage time is earlier than that listed for the sun, the moon is "ahead" of the sun, meaning to the west of it. With practice, the rising, meridian passage, and setting times of the moon relative to the sun, will give you a fairly good idea of the moon's location.

Referring to Table 3-3, we see that the moon is typically best positioned for daytime sights with the sun during moon ages 6-8 and 21-23 days, or about one week of each month. In practice, though, these sights might be available for longer periods, up to almost two weeks in some cases (ages 4-10 and 19-25), depending on your latitude and the moon's declination. In some of these extended cases, however, the optimum sight times might be so close to twilight that there is no virtue in the sun-moon fix when you have a star fix available within a couple of hours. Procedures for picking the best sight times, and why it is important, are discussed in Chapter 6.

Table 3-3 also tells when you might combine the moon with star sights. Again, just use it as a guide and then make more specific choices from your specific circumstances as illustrated later on. It could be, for example, that the moon is indeed visible for evening twilight sights on the day (age) that the Table predicts, but for your location and date it is too low for an accurate sight.

In some circumstances you can take star sights at night by a moonlit horizon. The accuracy will not be the best, but it is possible when you have a bright moon that is fairly high in the sky. The trick is to pick the 3 optimum stars (as explained in Chapter 5) and then hope that the errors will be about the same in each of the sights. Any one sight will not be very accurate, but if the errors for each star are roughly the same your fix will be a good one. The errors are more likely to be similar if the horizon is similar in all directions. Your 3 LOPs will make a fairly large triangle, but its center might be a reasonable fix. The procedure is not recommended for routine use.

3.5 Time and Time Keeping in Navigation

To use the Star Finder you need to know how to figure universal coordinated time (UTC) from your watch time (WT), and vice versa. You will also need to figure UTC and WT from local mean time (LMT), which is a special time concept used in navigation and astronomy. For practical application this is all we need to know of time, but candidates for Coast Guard exams must also be familiar with zone time (ZT) and chronometer time (CT).

Assuming your watch has no error, the difference between your WT and UTC is a fixed, whole number of hours. Eastern standard time, for example, is exactly 5 hours behind UTC. If your watch reads 1026 EST, it is 1526 UTC. It is important to remember that if your watch is set to EST, this 5-hour difference applies regardless of where you are in the world when you read this watch. The fact that your watch may not be set to the proper time zone for a particular region when you are traveling does not affect how you get UTC from that watch.

The number of hours that a watch differs from UTC is called the Zone Description (ZD) of the watch—or more precisely, the ZD of the time zone your watch is set to. The ZD can be (+) or (–) depending on whether the WT is behind or ahead of UTC. Standard time zones in western longitudes have a positive (+) ZD, while eastern longitudes have negative (–) ZD. Rephrasing the caution from the previous paragraph: It is the "ZD of the watch" that we need to know, not the ZD of the place we happen to be. In equation form, we find UTC by applying the ZD of the watch to the WT:

$$UTC = WT + ZD,$$

where ZD can be a positive or negative number.

You can find the ZD of your watch's time zone (standard or daylight saving) from the *Nautical Almanac*—it lists these for all parts of the world and tells whether and when Daylight Saving Time is used. Or, better still, you can receive UTC directly from the National Bureau of Standards WWV radio broadcasts. These can be received on nearly any shortwave radio (at 10, 15, or 20 Mhz) or by a telephone recording from the National Bureau of Standards in Fort Collins, Colorado. The telephone number is (303) 499-7111. This is not a toll free call; it is available 24 hours a day. On radio or telephone the time is announced only on the exact minute; ticks at one-second intervals are heard in between, except during the weather broadcasts and special messages. The 29th tick is skipped, so you can use the 30th as a midway marker.

Larger vessels at sea typically set ship's clocks by a time system that is slightly different from the standard time Zone system used on land. Ships and ship's navigators (and Coast Guard examinations) use zone time (ZT). Zone time is a simplified version of standard time, wherein the time zones change at specific longitudes. Each zone time zone is 15° of longitude wide, centered on each meridian around the world that is an even multiple of 15°: 0° (ZD = 0), 15° E (ZD = –1 hr), 15° W (ZD= +1 hr), 30°E (ZD= –2 hr), 30°W (ZD= +2 hr), and so forth. The central longitude of each zone is called its standard meridian; the boundaries of each zone are at longitudes of 7° 30' to either side of the standard meridian. For example: ZD = +8 labels the time zone that extends from longitude 112° 30' W to 127° 30' W. Its standard meridian is at $8 \times 15° = 120°$ W. Daylight saving time is never used in the zone time system.

Note that zone time boundaries occur at the logical intervals since the sun circles the earth at a rate of 15° of longitude each hour (360°/24 hr = 15°/hr). If you want noon to always be at about 1200 on your watch as you sail around the world, you must change the time zone of your watch by 1 hour each time you cross 15° of longitude.

The zone time and standard time systems are essentially the same except for way the time zone boundaries are specified. Zone time boundaries occur at specific longitudes regardless of where or what these longitude lines cross. Standard time zone boundaries, on the other hand, follow as closely as possible the zone time boundaries but they jog and curve to fit the convenience of local governments and geography. In most cases you can guess the ZD of a particular standard time zone from the longitude of the region using the zone time prescription, but in other cases you cannot—especially during the summer half of the year, since not all areas choose to switch to daylight saving time, which changes the ZD by 1 hour.

If you are navigating by ZT—which typically means changing the ship's clocks by one hour whenever a ZT boundary is crossed—you must figure the appropriate ZD from your longitude before you can determine UTC. To do this you round off your DR longitude to the nearest whole degree, divide by 15, and then round off the answer to the nearest whole hour. In west longitudes the ZD is (+); in east longitudes it is (–). For example, if I am looking at the stars at 0545 ZT from a DR longitude of 159° 23' W, the ZD is 159/15 = +10.6, rounded to +11 hr, and the UTC = ZT + ZD = 0545 + 1100 = 1645 UTC.

The life of the navigator is much simpler if you don't have to change your watch—meaning you use watch time for navigation instead of zone time. Example: the ZD of my watch is +7 hr—it has been from the beginning of this cruise and it will be until I get there. My watch reads 1315, so UTC = 2015. I don't care what my longitude is.

But there are limits to this strict use of WT. If my WT gets too far out of phase with the local ZT it will be an awkward time piece for daily functions. A compromise is usually in order, such as changing the ZD of your watch after each ocean crossing. This will typically keep your watch time to within 3 hours or so of the local zone time. My own preference is for WT whenever possible, but I still want morning times to be small numbers and evening times to be big numbers.

Zone time is more awkward for navigation than watch time, but it does have some virtue: it helps normalize your daily life when you live and work for long periods of time at sea, crossing large spans of longitude. Mealtime and overtime scheduling, for example, are much better handled by a zone time system.

Chronometer time (CT) is a time system used on some Coast Guard exams. It is UTC kept on a 12-hour clock without specifying AM or PM. When the time of some event is given as, say, 0455 CT, your first job is to figure out what time of day they are talking about (0455 UTC or 1645 UTC), using related information in the problem. It is an exercise that does indeed test your knowledge of time in navigation, but this method of time keeping has no place in practical navigation. If CT were specified on a 24-hr watch face, it would be the same as WT with ZD = 0 hr, which then makes it a reasonable time concept—the watch time of a watch set to UTC.

We now come to the topic of local mean time. This is a special—and indeed important—time concept that lets you figure the time your watch will read when some celestial event occurs, regardless of where you happen to be located within a particular time zone. Consider, for example, the celestial event to be the meridian passage of the sun—in northern latitudes, the time the sun is due south of you, at its maximum height in the sky. The sun circles the earth at 15° of longitude each hour. If we had a string of navigators stretched east-to-west across a time zone, with their watches all set to the same time, they would each record a different time for meridian passage. If the guy at the central meridian recorded 1200 for meridian passage, the guy at the eastern boundary would have recorded 1130, and the guy at the western boundary would record that the sun passed him at 1230.

Clearly the precise time of meridian passage depends on your longitude even if your watch time is set to the proper zone time. Also it should be clear that if the Almanac is to tell us the time of meridian passage, sunset, twilight, etc., it cannot list single times that will be valid for all longitudes within a specific time zone, even if everyone in the time zone has their watches set to the proper ZD for that time zone. This problem is solved by using local mean time (LMT). All of these special times listed in the Almanac are given in terms of LMT. We must then figure the UTC, ZT, or WT of the event using the listed LMT and our DR longitude.

There are several ways to think of this LMT. The ZT of some event, say sunrise, will be exactly the same as the listed LMT of the sunrise if you happen to be exactly in the middle of a zone time zone, on a standard time meridian—your longitude an exact multiple of 15°. At longitudes east of the central meridian, the ZT of sunrise will be earlier than the listed LMT; west of the central meridian it will be later. But this reasoning does not work if you are using WT and your watch is not set to the proper ZT of your location. There is another, better way to think of LMT that will work for WT or ZT, and it is the recommended procedure for all applications.

Since the time we need to know for any celestial sight (or use of the Star Finder) is UTC, the safest procedure is to figure UTC directly from LMT, and then from there figure ZT or WT if we need it. Zone time at the Greenwich meridian is the same as UTC, so the UTC of sunrise is the same as the listed LMT of sunrise if you are at longitude 0°. West of Greenwich the UTC of sunrise will be later than the listed LMT by 1 hour for every 15° of longitude you are west of Greenwich. East of Greenwich it will be earlier by the same amount.

A typical problem we have with the Star Finder is to figure the UTC of twilight for a particular date, latitude, and longitude. The Almanac tells us the LMT of twilight for the date and latitude, and we then use our longitude to figure how much the UTC will differ from the LMT listed. It is a simple matter to convert the LMT to UTC, and, if we like, from there to WT or ZT. To do this we use the Arc to Time Table in the *Nautical Almanac*, which converts longitudes to times at the rate of 15° per hour.

Procedure: Convert your DR longitude (DR-Lon) to time using the Arc to Time Table in the *Nautical Almanac*. The longitude to use here is simply your best estimate of your longitude at the time you plan to do celestial sights or look at the stars. For most applications it does not matter if this longitude is not exact, since we don't know to the minute when we will actually do the sights or look at the stars. If your longitude is not exact, the UTC you figure will not be exact. A 1° longitude error will cause a 4 minute time error, so even if your longitude is off by a couple degrees (which is unlikely) you will get the time right to within 10 minutes or so.

In equation form:

$$UTC = LMT + DR\text{-}Lon(W),$$
or
$$UTC = LMT - DR\text{-}Lon(E),$$

where the longitudes must be converted to time using the Arc to time Table.

Once you have UTC you can look up the other numbers you need from the Almanac to set up the Star Finder. If you want to know what your watch will read at this UTC, just do backwards whatever you do to get UTC from WT. If I get UTC by adding 7 hours to my WT, then to find WT from UTC, I subtract 7 hours. The result is the WT I should do the sights. If you are using ZT, then use the ZD of your location for this conversion, rather than the ZD of the watch—although they are likely to be the same in this case.

Numerical examples in following chapters illustrate this procedure in specific circumstances.

Example 3-2
Finding The UTC and WT of Twilight

The date is July 14th, 1982; our DR position is 38° 20' N, 67° 50' W. Our watch is set to ZD = +4 hr. What is the UTC, ZT, and WT of evening civil twilight?

From the Almanac daily pages (see Appendix) we learn that evening civil twilight at latitude 35° N is 1944 LMT and at latitude 40° N is 2001 LMT. We are about halfway between these two, so we take the time about halfway between the two listed, which we can get by averaging the two: (19h 44m + 19h 61m)/2 = 19h 52.5m which we round to 1953 LMT.

Note the time averaging is simplified by rewriting the hours parts with the same hours, and for general use of the Star Finder we only need to know times accurate to the minute, rounding off the seconds whenever they appear. Also, the single times listed on the 3-day daily pages are for the middle day of each page, the 13th in this example. But these times change only slowly from day to day, so you can use the listed time for any day on the page, 12th, 13th, or 14th.

To find UTC we need to convert our DR-Lon to time using the Arc to Time Table in the Almanac: 67° 50' = 4h 31m 20s, which we round to 0431. And now UTC = LMT + DR-Lon(W) = 1953 + 0431 = 23h 84m UTC on July 14th = 24h 24m on July 14th = 0024 on July 15th UTC. This is the UTC we would use to look up the GHAs and declinations from the Almanac.

To find WT from UTC, we just reverse the procedure for finding UTC from WT, recalling that the ZD of the watch is defined as the number of hours you must add to WT to get UTC. In this example, ZD = +4 hr, so WT = UTC − 4 hr = 0024 − 0400 = 2424 − 0400 = 2024 WT on July 14th.

To find ZT we must first figure the ZD of our longitude, by rounding the longitude to the nearest degree, dividing by 15, and then rounding the result to the nearest hour: (67° 50')/15 or 68°/15 = 4.53 or +5 hr. The "+" sign is used because we are in west longitudes. We find ZT just as we found WT: ZT = UTC − ZD = 0024 − 0500 = 2424 − 0500 = 1924 ZT on July 14th.

Note that in this example our navigation watch (ZD = +4 hr) was not set to the "proper" ZD of our location, so ZT and WT were not the same. But this does not affect our celestial navigation, it only determines how we will label our chart work and log books. The critical time is the UTC of the event. Here the event is evening civil twilight and it occurs at 0024 UTC July 15th, as viewed from this location. Every ship in the neighborhood will agree with this, regardless of what time system they use to keep their navigation records.

Example 3-3
More on Finding UTC and WT.

This is a similar example, but now from a different location and event. The date is July 14th, 1982; our DR position is 17° 44' S, 93° 25' E. Our watch is set to ZD = –6 hr. What is the UTC, ZT, and WT of meridian passage of the sun (also called local apparent noon or LAN)?

From the Almanac (see Appendix) we learn that meridian passage on July 14th (for all latitudes) is 1206 LMT. Our DR-Lon converted to time is: 93° 25' = 6h 13m 40s, which we round to 0614. And now UTC = LMT – DR-Lon(E) = 1206 – 0614 = 11h 66m – 06h 14m = 05h 52m = 0552 UTC on July 14th. This is the UTC we would use to look up the GHAs and declinations from the Almanac.

To find WT from UTC, we reverse the procedure for finding UTC from WT: In this example, ZD = –6 hr, so WT = UTC + 6 hr = 0552 + 0600 = 1152 WT on July 14th.

To find ZT we first figure the ZD of our longitude: (93° 25')/15 or 93°/15 = 6.20 or –6 hr. The "–" sign is used because we are in east longitudes. We find ZT just as we found WT: ZT = UTC + ZD = 0552 + 0600 = 1152 ZT on July 14th. In this case ZT and WT are the same, and both agree that noon will be at about 1200.

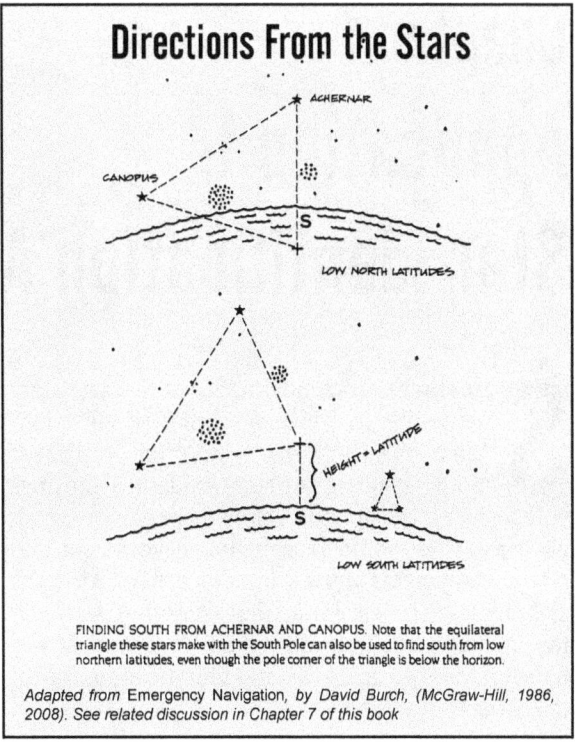

FINDING SOUTH FROM ACHERNAR AND CANOPUS. Note that the equilateral triangle these stars make with the South Pole can also be used to find south from low northern latitudes, even though the pole corner of the triangle is below the horizon.

Adapted from Emergency Navigation, *by David Burch, (McGraw-Hill, 1986, 2008). See related discussion in Chapter 7 of this book*

Chapter 4.
Application I:
Star Identification After the Sight

Sometimes on cloudy nights only a few stars are visible in isolated patches of clear sky. When this happens, you often can't see enough neighboring stars to identify the ones you do see. If the horizon below these stars is good, however, there is no reason not to go ahead and take sextant sights of these unknown stars. After the sights, you can figure out what stars they were, do the sight reductions, and get a fix. The simplest and most reliable way to do this star identification is with the 2102-D Star Finder.

To do this you only need to do one extra thing at the time of the sight that is not part of a regular star sight: you need to measure the true bearing of the star at the time of the sight. You can get this from a magnetic bearing using a handheld compass or, in some circumstances, by simply sighting the star over the steering compass. The magnetic bearing can then be converted to a true bearing using the local magnetic variation. It is not necessary to know this bearing accurate to the degree, but the better you know it the easier the identification might be. In any event, it is good practice to take the star's bearing as carefully as you can.

Note that since you won't know what stars you are sighting, you won't have precomputed the star's height and bearing—the standard way of preparing for star sights, as discussed later in Chapter 5. In this circumstance, then, it pays to remember the trick of inverting the sextant to bring the horizon up to the star.

Without precomputation, an inverted sextant is, without a doubt, the best way to do the sights: Set the sextant to 0" 0', invert the sextant, hold it in your left hand, and, looking up toward the star, sight it in the direct-view side of the horizon glass. Reach over the top of the sextant and, keeping the star in direct view, move the index arm forward to bring the horizon up to the star. Once star and horizon are both in view, turn the sextant back over to its normal position, and complete the sight in the normal way.

Don't try to precisely align the star with the horizon while the sextant is upside down; this is difficult and unnecessary. Just get them both in view. If you lose the star while bringing the horizon up, it is best to start over again at 0" 0', rather than hunt around for the star with the sextant set at some intermediate angle.

4.1 Identifying Navigational Stars

STEP 1. Do the star sight as described above to get the sextant height (Hs) at a known watch time (WT). Then take the magnetic bearing to the star as carefully as you can, and record it with Hs and WT.

STEP 2. Convert the star's magnetic bearing to a true bearing by applying the local magnetic variation. Magnetic variation is given on Pilot Charts or any other chart of the waters you are in.

STEP 3. Convert WT to UTC by correcting for watch error (if any) and ZD—the latter is just the number of hours between the time zone of your watch and UTC. Then look up the Greenwich Hour Angle (GHA) of Aries in the *Nautical Almanac* for the UTC of the sight that you just figured. Next use your DR longitude (DR-Lon) to find the Local Hour Angle (LHA) of Aries. The formula depends on whether your DR-Lon is East or West.

LHA Aries = GHA Aries − DR-Lon(W)
or
LHA Aries = GHA Aries + DR-Lon(E).

If the result is negative, add 360°; if the result is over 360°, subtract 360°. LHA Aries should end up as a positive number, less than 360°.

STEP 4. Look through the set of blue templates and find the one for the latitude that is closest to your DR-Lat—in other words, closest to the latitude you think you were at when you did the sight. Place this blue template on the pin at the center of the white disk. Be sure to have the north side of the blue template on the north face of the white disk if you are in northern latitudes. As you rotate the blue template on the white disk, the blue arrow will scan along the black scale on the rim of the white disk. Rotate the blue template until the blue arrow points to the LHA Aries you just figured.

This operation is effectively rotating the sky over your head to set it at the proper orientation for your position at that time.

Note that the LHA Aries you figure from the Almanac and your longitude will have degrees and minutes, but the rim scale is only marked to the nearest half degree (0.5° or 30'). You can round off LHA Aries to the nearest 30' or simply interpolate the minutes as you set it. In the examples to follow, we usually convert LHA Aries to decimal degrees to help with this interpolation, but this is a fine point. In practice it won't matter how you do it.

STEP 5. Hold the template in place (arrow pointing to LHA Aries on the rim scale) and find your star by going around the edge of the diagram on the blue template until you find the true bearing of the star you observed. Then go along that bearing line, toward the center of the diagram, until you reach the point on the height scale that matches the height of the star (Hs) that you observed. If the star you observed was one of the 57 navigational stars listed on the daily pages, you will find it at the bearing and height you observed—or near there, within a few degrees. If this happens, you have identified your star. You can now look up its SHA and Dec in the *Nautical Almanac* and complete the sight reduction in the normal manner.

Follow through the following examples of the procedure with your Star Finder. Note that this first example (4-1) is more of a practice exercise than a practical situation, since it is unlikely that you could observe so many unknown stars, in so many different directions during one sight session. This is especially the case if you have prepared your star sights by precomputation as explained in Chapter 5.

Example 4-1 Star ID: Navigational Stars

On July 13th, 1982, at DR position 23° 35' N, 149° 15' W, the following sights of unknown stars were taken during a cloudy evening twilight:

Brightness		Sextant Height
Star 1	Faint	47°57.1'
Star 2	Bright	49°14.6'
Star 3	Faint	41°36.2'
Star 4	Bright	49°21.6'
Watch time		Magnetic Bearing
Star 1	19h 12m 25s	085 M
Star 2	19h 22m 32s	200 M
Star 3	19h 28m 35s	150 M
Star 4	19h 36m 11s	220 M

The Watch Error is 5s Fast; the ZD of the watch is +10 hr; and the local magnetic variation is 12° East. Identify the stars.

Star 1: UTC of sight = 19h 12m 25s − 5s + 10h = 29h 12m 20s on July 13th = 05h 12m 20s on July 14th. From the Almanac (see Appendix), at this UTC, the GHA Aries = 09° 49.8', so the LHA Aries = 09° 49.8' − 149° 15.0' (+ 359° 60') = 220° 34.8' = 220.6°. The true bearing of the star = 085° M + 12° E = 097° T ("Correcting add East").

The closest blue template to DR-Lat 23° 35' is 25°. Place the north side of this template on the North face of white disk, and set the blue arrow to 220.6° on the rim scale, as shown in Figure 4-1. Find the bearing 097°, halfway between the 095° and 100° lines, and follow this line toward the center of the diagram to the observed height of about 48°. And there is your star, Rasalhague, a magnitude two navigational star.

The Star Finder Book

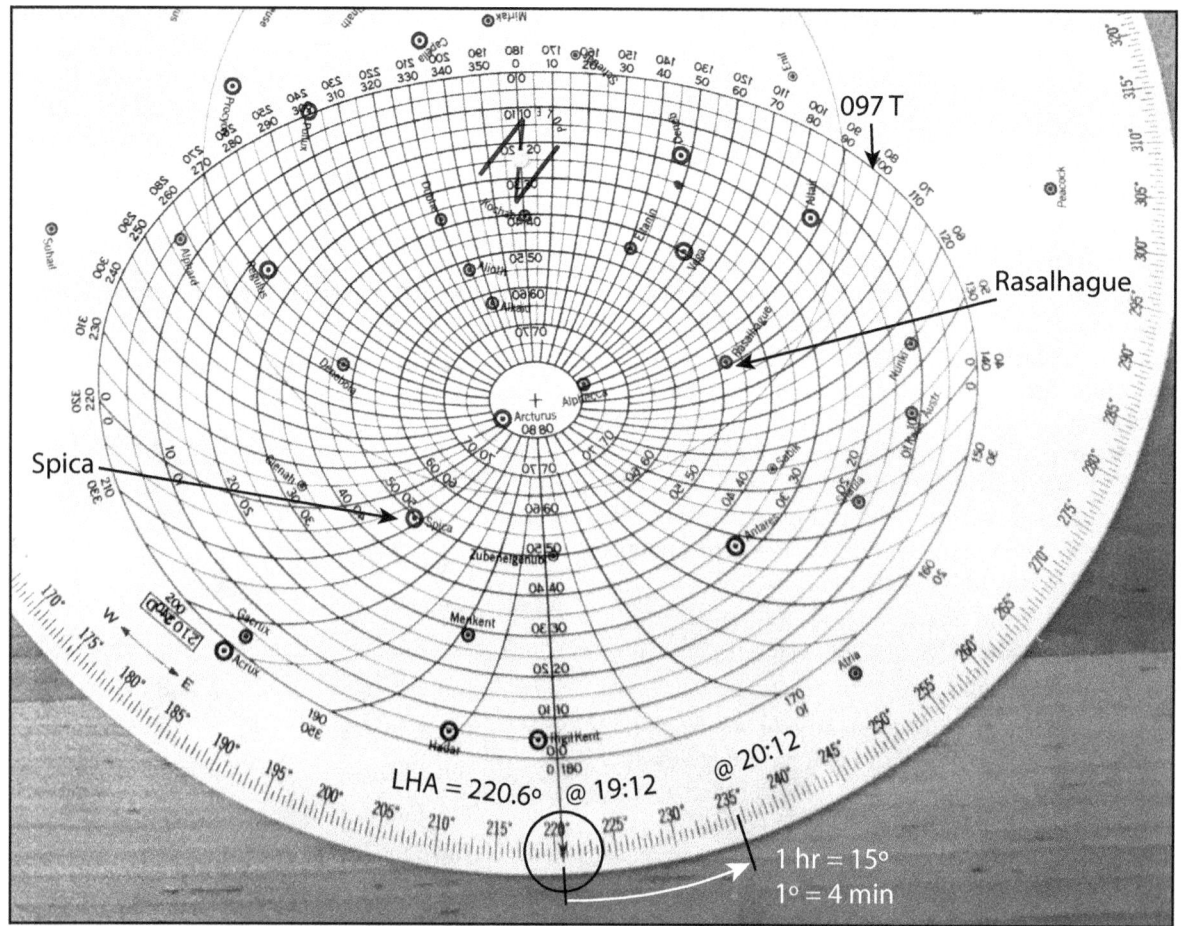

Figure 4-1. *The 25° N template set at LHA Aries = 220.6° to identify star Rasalhague in Example 4-1. The star was observed at bearing 097° T and height 48°. The rim scale is marked with the watch time of the sight.*

Star 2: We could do this one the same way—that is, convert WT to UTC, look up GHA Aries and then figure the proper LHA Aries for this sight—but there is an easier way.

We already know that 220.6° on the rim scale corresponds to the watch time 19h 12m 25s of the first sight. The second sight at WT 19h 22m 32s is 10m 7s later. Since the rim scale corresponds to time at the rate of 15° per hour, we can easily figure how much to rotate the blue arrow without going through the tables and arithmetic again: 15° per hr is the same as 1° per 4m, so in 10m 7s the scale would rotate about 2.5°. The proper setting for the second sight is 220.6 + 2.5 = 223.1°. Another approach is to label the rim scale with watch time as shown in. Figure 4-1, and then simply estimate a 10 minute interval along this scale.

The true bearing of Star 2 is 200° M + 12° E = 212° T. Set the blue arrow to 223.1°, go to bearing line 212°, and follow it to Hs = 49°. Your star is Spica, a magnitude one navigational star. You can see this from Figure 4-1 if you imagine the slight rotation of the blue template.

Remember, we can't hope to always find our star at the precise height and bearing we measured. We are not using the blue template for our exact latitude, only the nearest, and our measured bearing could easily be off a few degrees. But with care we

24

STARS, 1982 JANUARY—JUNE

Mag.	Name and Number		S.H.A.						Dec.					
			JAN.	FEB.	MAR.	APR.	MAY	JUNE		JAN.	FEB.	MAR.	APR.	MAY	JUNE
3.0	δ Ophiuchi	116	39.4	39.2	39.0	38.8	38.7	38.6	S. 3	38.9	39.0	39.0	39.0	39.0	38.9
2.8	β Scorpii	118	54.6	54.4	54.2	54.0	53.8	53.8	S. 19	45.3	45.3	45.4	45.4	45.5	45.5
2.5	δ Scorpii	120	11.5	11.2	11.0	10.8	10.6	10.6	S. 22	34.1	34.2	34.2	34.3	34.3	34.3
3.0	π Scorpii	120	34.1	33.8	33.6	33.4	33.2	33.2	S. 26	03.6	03.7	03.7	03.8	03.8	03.8
3.0	β Trianguli Aust.	121	37.6	37.1	36.7	36.3	36.0	36.0	S. 63	22.3	22.3	22.3	22.4	22.6	22.7
2.8	α Serpentis	124	09.7	09.5	09.3	09.1	09.0	08.9	N. 6	28.9	28.8	28.8	28.8	28.9	28.9
2.3	α Coronæ Bor. 41	126	31.5	31.3	31.1	30.9	30.8	30.8	N. 26	46.4	46.3	46.3	46.3	46.4	46.6
3.0	γ Lupi	126	31.4	31.1	30.9	30.6	30.5	30.4	S. 41	06.2	06.2	06.3	06.4	06.5	06.5

STARS, 1982 JULY—DECEMBER

Mag.	Name and Number		S.H.A.						Dec.					
			JULY	AUG.	SEPT.	OCT.	NOV.	DEC.		JULY	AUG.	SEPT.	OCT.	NOV.	DEC.
3.0	δ Ophiuchi	116	38.6	38.7	38.8	38.9	38.9	38.9	S. 3	38.7	38.9	38.9	38.9	38.9	39.0
2.8	β Scorpii	118	53.8	53.9	54.0	54.1	54.1	54.0	S. 19	45.5	45.5	45.4	45.4	45.4	45.4
2.5	Dschubba	120	10.6	10.7	10.8	10.9	10.9	10.8	S. 22	34.3	34.3	34.3	34.3	34.3	34.3
3.0	π Scorpii	120	33.2	33.3	33.4	33.5	33.4	33.4	S. 26	03.9	03.9	03.8	03.8	03.8	03.8
3.0	β Trianguli Aust.	121	36.1	36.3	36.6	36.8	36.9	36.7	S. 63	22.8	22.9	22.8	22.8	22.6	22.5
2.8	α Serpentis	124	09.0	09.1	09.2	09.3	09.3	09.2	N. 6	29.0	29.0	29.0	29.0	28.9	28.8
2.3	Alphecca 41	126	30.8	30.9	31.1	31.2	31.2	31.1	N. 26	46.6	46.7	46.7	46.6	46.5	46.3
3.0	γ Lupi	126	30.5	30.6	30.8	30.9	30.9	30.7	S. 41	06.6	06.6	06.6	06.5	06.5	06.4

Figure 4-2. *Selection from the extra star list at the back of the* Nautical Almanac. *In Example 4-2 the approximate values of SHA = 121° and Dec = S 22° for Star 3 most closely match those of star Dschubba. Hence the unknown Star 3 was Dschubba.*

should find our star, if it's a navigational one, within a few degrees of our estimate.

Star 3: The third sight was taken 6 minutes after the second, which corresponds to a further blue arrow rotation of 1.5°. The rim scale for this sight is 223.1° + 1.5° = 224.6°, and the true bearing is 150° + 12° = 162°. Set the blue arrow and go up the 162° bearing to the observed height of about 42°. And what do you find? Nothing. (See your own Star Finder, or imagine the further slight rotation of the blue template shown in Figure 4-1.) The closest star is Antares, but it is off by 6° or 7° in both height and bearing. Furthermore, Antares is bright and red (Table 3-2). The star was not Antares. It was not a navigational star at all.

The rest of the instructions will tell us what to do at this point. By the way, if you work out the Spica-Rasalhague fix, you will find the DR position was accurate (Index Correction = 0.0′; Height of Eye = 9 ft).

4.2 Identifying Other Stars

We will continue on with the STEP sequence since you would have to do the earlier ones to get to this point.

STEP 6. If there is no star on the white disk at the height and bearing you observed, the star was not a navigational star; but this does not mean we can't use it, whichever it is, for navigation. Holding the blue template fixed, lift one corner of it and place a pencil mark on the white disk at the star's observed height and bearing. Let the template back down and check the height, bearing, and LHA Aries scales to verify that you put the mark at the right spot. Now we have our star on the white disk at the right spot, it just doesn't have its name beside it.

Our task now is to figure the Dec and SHA of this spot, which will be the approximate values for the unknown star. Using these approximate values, we then check the list of extra stars at the back of the

The Star Finder Book

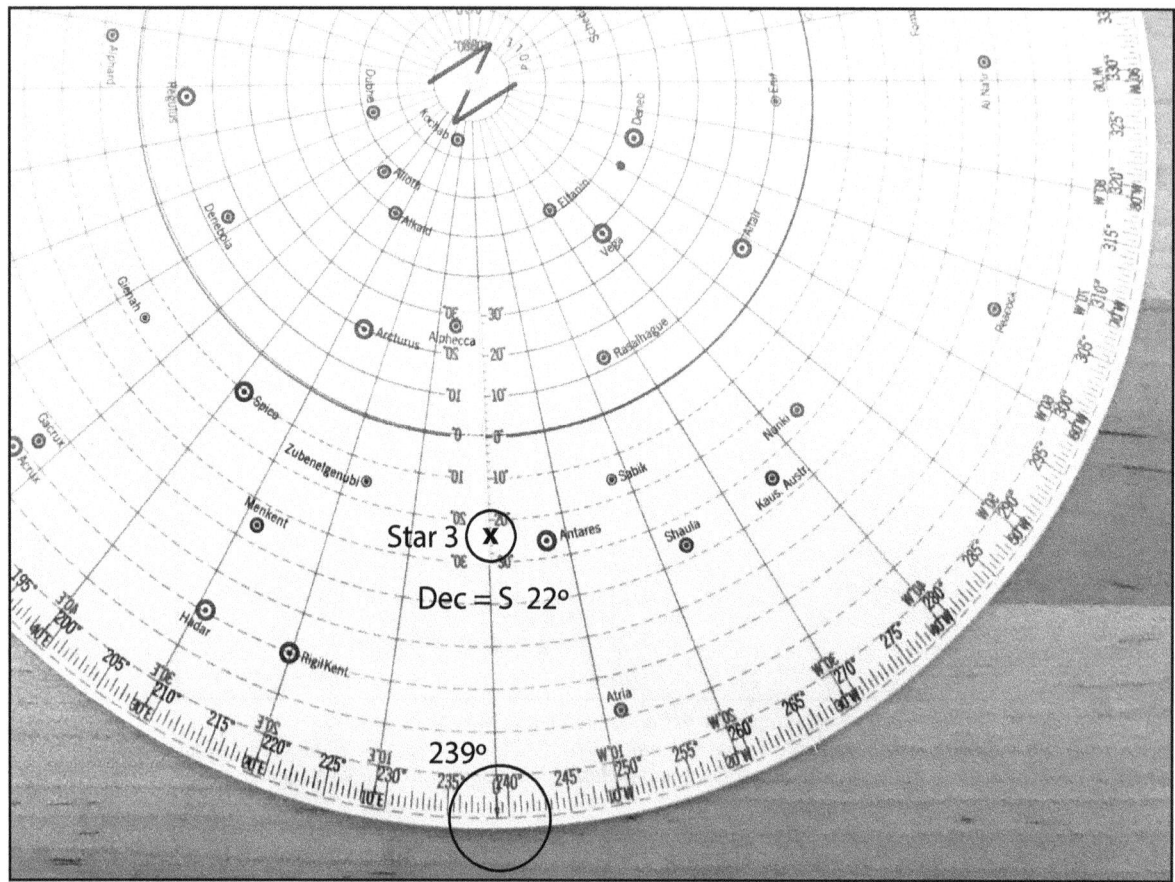

Figure 4-3. *The red template set at the unknown Star 3 in Examples 4-2. It was observed bearing 162° at a height of about 42°, and its position on the white disk was marked using the blue template shown in Figure 4-1, as explained in Example 4-2. With the red template set as shown, read Star 3's Dec as about S 22° and its rim scale position as 239°. Its SHA = 360° − rim-scale reading = 360° − 239° = 121°.*

Nautical Almanac to find the exact values needed for the sight reduction.

STEP 7. With the mark on the white disk, remove the blue template and replace it with the special template with red printing. Use the south side of the red template, as it shows the Dec scale more clearly. The red template has one radial line with an arrow on it and an open slot near its middle. Rotate the template until this radial line crosses your mark, and read the Dec of the star from the scale on the radial line. Part of the Dec scale is printed on this radial line; the rest of the scale is on the line opposite it. You will have to interpolate this value since lines are only given every 10° but an accuracy of 2 or 3° should be good enough.

Now figure out if the Dec is north or south. The Dec will be north if you are on the north side of the white disk (sailing in northern latitudes) and the mark is in the center part of the disk, in the solid red lines. If your mark is on the outer part of the disk in the dashed lines, the Dec is south. The opposite is true (dashed is north, solid is south) if you are in southern latitudes using the south side of the white disk. If you are uncertain about this, use the Almanac to look up the Dec of a nearby named star to see if it is north or south.

STEP 8. Now look at the red arrow pointing to the black scale on the rim of the white disk—the scale we used earlier for LHA Aries. Take the reading on this scale and subtract it from 360°; the result is the SHA of the star.

SHA = 360' − (white-disk rim scale reading).

STEP 9. You now have approximate values of the Dec and SHA of the star. Go to the extra star list at the back of the *Nautical Almanac* and find the stars that have the approximate SHA you just figured. This star list is organized according to SHA, which makes this step easy. Of these listed stars with about the right SHA, only one will have the right, or nearly right, declination. This is your star. Record its exact SHA and Dec for the appropriate month, and proceed with a normal star sight reduction. You have identified your star.

A section of this extra star list is shown in Figure 4-2. The SHAs and Decs are listed only monthly for the stars here, as opposed to every 3 days in the daily pages, but this difference is insignificant. Also note that for 6 months of the year the stars are called by their Greek-letter names, and for the other 6 months their proper names are given, if they have one. The same star goes across both pages, but it might be listed with two different names.

Example 4-2
Star ID: Non-Navigational Stars

Star 3 of Example 4-1 was observed at Hs = 41° 36.2' at a true bearing of 162° T at UTC = 05h 28m 30s, July 14, 1982. The DR position of observation was 23° 35' N, 149° 15' W. We will skip the time conversions and magnetic to true conversions for the rest of this example.

From the Almanac, at that UTC, the GHA Aries = 13° 53.0', and the LHA Aries = 13° 53.0' − (149° 15.0') + (360°) = 224° 38.0' = 224.6°. With the arrow of the 25° blue template set at 224.6° on the rim scale of the white disk, we find no star at a height near 42° and bearing near 162°.

Reach under the blue template and place a mark on the white disk as close as you can to height 41.5° and bearing 162°. Double check that height, bearing, and rim scale are correct, then replace blue template with the red one and align it with this mark as shown in Figure 4-3.

Read the approximate Dec of the star from the circular scales on the red template, about S 22°– south because it is on the dashed scales. And read the rim scale on the white disk; it should read 239°. The star's approximate SHA = 360° − rim scale reading = 360° − 239° = 121°.

We now turn to the star list at the back of the *Nautical Almanac* to find a star with SHA about 121° and Dec of about S 22°. From this list (Figure 4-2) we find that the star was Dschubba, a faint magnitude two star in Scorpio; its exact values of SHA and declination for July are 120° 10.6' and S 22° 34.3'.

An alternative approach to reading the approximate values of Dec and SHA of a star or planet not printed on the white disk is this: instead of placing a mark on the white disk and removing the blue template, hold the blue template in place on the white disk and put the red template over it. Align the slotted radial line of the red one over the proper, though unmarked, position on the blue scales—at the height and bearing you observed. Then, as before, read the Dec from the red template and the value from the rim scale on the white disk (SHA = 360' − white-disk rim scale reading). Some navigators prefer this method. Try both, and pick the one you like.

4.3 Identifying Planets

And now the final step in the sequence of identifying unknown objects: STEP 10. If no star listed in the extra star list at the back of the Almanac matches the SHA and Dec you found from the Star Finder, and no mistake was made, then you probably have a planet. The declinations of the planets are listed on the daily pages of the *Nautical Almanac* for each hour. The SHAs of the planets are also listed on the daily pages, at the bottom of the star column, (see Appendix). Planet SHAs are only listed once every three days since they don't change very rapidly. The SHAs of planets are only provided for planet identification, not for direct navigational applications, which is another reason their precise hourly values are not given.

STARS, 1982 JULY—DECEMBER

Mag.	Name and Number		S.H.A.						Dec.					
			JULY	AUG.	SEPT.	OCT.	NOV.	DEC.	JULY	AUG.	SEPT.	OCT.	NOV.	DEC.
2·8	η Bootis		151	32·4	32·5	32·6	32·6	32·6	32·4	N. 18 29·4	29·4	29·3	29·3	29·1	29·0
1·9	Alkaid	34	153	17·4	17·6	17·7	17·8	17·7	17·5	N. 49 24·4	24·4	24·3	24·1	24·0	23·8
2·6	ε Centauri		155	18·6	18·8	19·0	18·9	18·6		S. 53 22·8	22·8	22·7	22·6	22·5	22·4
1·2	Spica	33	158	56·2	56·3	56·3	56·4	56·3	56·1	S. 11 04·1	04·1	04·1	04·1	04·1	04·2
2·2	Mizar		159	11·9	12·1	12·2	12·3	12·2	11·9	N. 55 01·4	01·4	01·2	01·1	00·9	00·7
2·9	ι Centauri		160	06·1	06·2	06·3	06·3	06·2	06·0	S. 36 37·3	37·2	37·2	37·1	37·0	37·0
3·0	ε Virginis		164	40·7	40·8	40·8	40·8	40·7	40·5	N. 11 03·4	03·4	03·4	03·3	03·2	03·1
2·9	Cor Caroli		166	12·1	12·3	12·3	12·3	12·2	12·0	N. 38 25·1	25·1	25·0	24·9	24·7	24·5
1·7	Alioth	32	166	41·4	41·6	41·7	41·7	41·5	41·2	N. 56 03·7	03·6	03·5	03·3	03·1	03·0
1·5	Mimosa		168	19·9	20·2	20·3	20·3	20·1	19·7	S. 59 35·8	35·7	35·6	35·4	35·4	35·3
2·9	γ Virginis		169	48·7	48·8	48·8	48·8	48·7	48·5	S. 1 21·1	21·0	21·0	21·1	21·1	21·2
2·4	Muhlifain		169	52·1	52·3	52·4	52·3	52·2	51·9	S. 48 51·9	51·9	51·8	51·7	51·6	51·6
2·9	α Muscæ		170	58·5	58·8	59·0	59·0	58·7	58·2	S. 69 02·5	02·5	02·3	02·2	02·1	02·1
2·8	β Corvi		171	38·3	38·4	38·5	38·4	38·3	38·0	S. 23 18·0	17·9	17·9	17·8	17·8	17·9
1·6	Gacrux	31	172	27·5	27·7	27·8	27·8	27·6	27·2	S. 57 01·1	01·0	00·9	00·8	00·7	00·7

Figure 4-4. *Selection from the extra star list at the back of the* Nautical Almanac. *None of the listed stars match the approximate values of SHA = 164° and Dec = S 3°, which were determined for Star 4 of Example 4-3. Next step is to check the planet on the daily pages.*

If the SHA and Dec of your spot match any of these planet data, then your "star" was that planet. Note the name of the planet, record hits exact values of GHA and Dec for the time of the sight, and proceed with a normal planet sight reduction.

If you can't find a match in Dec and SHA with the navigational stars, the extra stars, or with the planets, then there is probably a mistake somewhere. Recheck all of your data and start again.

Note: An early hint that a non-navigational "star" might be a planet—besides the fact that it was bright and not a navigational star—is the value of its declination. To be a planet, the mark you plotted on the white disk must lie within the open slot on the red template. The declinations of all planets lie within that range. If the object was bright and in the slot it is a good candidate for a planet. Stars, of course, also have declinations in that range, so this is not positive proof, just a hint. Also, remember that Mars and Saturn do not always appear notably bright. On the other hand, if the declination is outside of the slot it cannot be a planet.

Example 4-3 Planet ID

Star 4 of Example 4-1 was observed at Hs = 49° 21.6' at a true bearing of 220 M + 12° E = 232° T at UTC = 05h 36m 06s, July 14, 1982. The DR position of observation was 23° 35' N, 149° 15' W.

From the Almanac, at that UTC, the GHA Aries = 15° 45.8', and the LHA Aries = 15° 45.8' − (149° 15.0') + (360°) = 226° 30.8' = 226.5°. With the arrow of the 25° blue template set at 226.5° on the rim scale of the white disk, we find no star at height near 49° at bearing near 232°.

Mark the observed position of the star on the white disk as explained in Example 4-2, and then replace the blue template with the red one to read the Dec and rim scale of the spot—exactly the same procedure as in Example 4-2. You should find the approximate value of Dec = about S 3° and the rim scale = about 195.5°. From the rim scale find the SHA = 360 − 195.5 = 164.5° = 164° 30'.

Checking the extra star list at the back of the Almanac (the relevant section is shown in Figure 4-4), we find no star with an SHA of about 164° that has

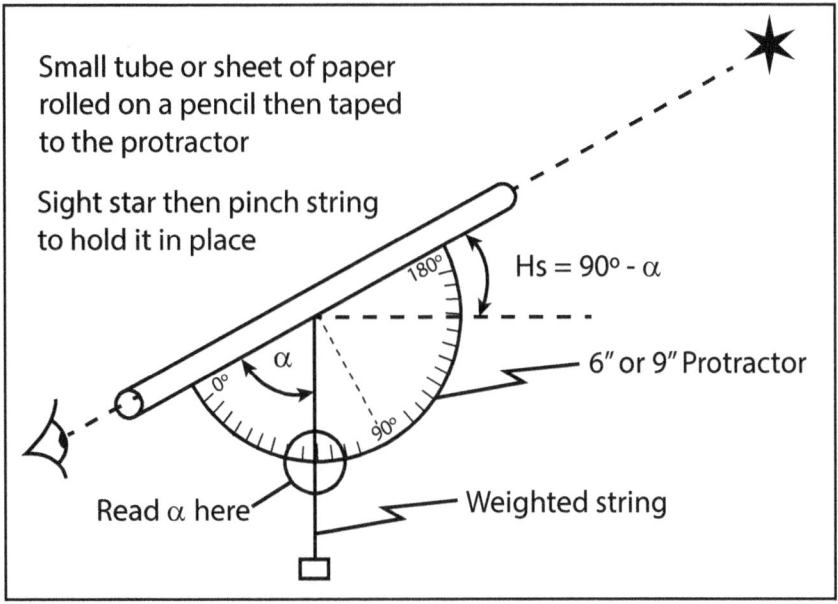

Figure 4-5. *A makeshift sextant for backyard practice on Star ID.*

a Dec of about S 3°. So Star 4 is not one of the extra stars. It was also not a navigational star or it would have been on the white disk. This leaves the planets.

Checking the planet data on the daily page for July 14th, (see Appendix), we see that "Star 4" was the planet Saturn.

There are further examples to practice with throughout this book, but once the paper work and Star Finder procedures are under control, the best thing to do by far, is start "real" practice, with real skies. You will need a current almanac and a small compass, but that is all—you can even practice without an almanac if you calculate GHA Aries using the formula given later in Chapter 7. You don't need a sextant to master the star ID phase of celestial navigation. Then just go out in the backyard and estimate the heights of a few bright objects and measure their bearings with the compass. Then identify them with the Star Finder.

Or work backwards from the procedures given in the next chapter: predict the heights and bearings of several objects and then go and look for them. This removes the problem of estimating their angular heights without a good horizon. You will soon find, however, that this approach is almost too easy. The other (measure the star then identify it) is more challenging, and consequently more rewarding. It is also the most important direct application of the Star Finder. Without a good horizon, you can measure the approximate angular heights of stars from land using a protractor as shown in Figure 4-5.

Chapter 5.
Application II:
Predicting Heights and Bearings

Routine star sights are much easier to do if you prepare a list of the heights and bearings of suitable stars before taking the sights. With this list, you then just set the sextant to the predicted height, look in the predicted direction, and the star will be in view near the horizon. The often troublesome tasks of identifying the star and getting it into view with the horizon are done automatically. This special preparation is called "precomputation."

The value of this procedure, however you choose to do it, cannot be over emphasized. Even if you know all the stars by heart, this is still the most efficient way to do star sights. Furthermore, and perhaps most important, if you can take the sights early in evening twilight by precomputing, the sea horizon is still sharp, which makes your sights not only easier, but also more accurate.

With the sextant set to the right height, you don't even have to worry about looking in the precisely right direction. Just scan the sky in the approximate direction, keeping the horizon in view, and your star will appear.

When anyone does this for the first time they are duly impressed. The appearance of a tiny white dot in the sextant telescope, all alone in a pale blue background, right where it's supposed to be, even though you can't see it with your naked eye, is almost magic. The exercise should be done early in one's study of celestial navigation; it is a quick and easy way to see that all this stuff really works.

This procedure works so well because, by precomputation, you are looking at the right part of the sky through the sextant telescope, which often allows you to see the star and take the sight before it is dark enough to see it with your unaided eye. This has the effect of extending the useful length of evening twilight, which for the navigator ends when it is too dark to see the horizon. This extra time is a big advantage near the equator where twilight times are short—where, as Kipling put it, "the dawn comes up like thunder, on the road to Mandalay."

Precomputation works especially well for the bright planets Jupiter and Venus. These you can find by precomputation the moment the sun goes down. Venus can sometimes even be found through a sextant telescope and used for navigation during the day while the sun is still above the horizon. Daytime sights of Venus are discussed in Chapter 6.

The 2102-D Star Finder will precompute the height and bearing of any celestial body, but we should mention from the beginning that there are other, sometimes preferable, ways to precompute stars. Stars (and stars alone) can be precomputed using Sight Reduction Tables for Air Navigation (Selected Stars) Pub. No. 249 Vol. 1. If you want to precompute stars alone—as opposed to picking an optimum star-planet combination—you may find Pub. 249 Vol. 1 a more convenient way to do it. The Tables list only 7 stars at a time, which may seem a limitation at first, but in practice it is not—in most cases there are not many more than about 7 of the 57 navigational stars suitable for sights at any one time.

The Pub. 249 Tables are typically faster than the Star Finder for precomputing stars—faster in the sense of producing the necessary accuracy in the shortest time. Accuracy in precomputation is important since the entire procedure is not much help if it is not accurate enough to get the star and horizon into view at the preset sextant height. If the seas are rough, hunting around for a star that is almost in view is not much different than hunting for one that is nowhere near in view.

On the other hand, star precomputation with the Star Finder has several advantages of its own. First, it is usually accurate enough to do the job on its own. And even if the accuracy might be in question, as when you happen to be located "halfway between two blue templates," you can always add the necessary accuracy by individual sight reductions of the stars you selected from the Star Finder. Second, the Star Finder takes up less space and weight than the Tables, which is an advantage on small boats. And third, the Star Finder lasts forever, whereas Pub. 249 Vol. 1 must be purchased every five years.

Most importantly, however, the Star Finder can be used for precomputing sun, moon, and planets, and Pub. 249 Vol. 1 cannot. If there is a bright planet visible at twilight, and you choose to use it—as you probably should—you will need both star and planet predictions. The Star Finder lets you do both jobs at once. Further advantages of the Star Finder for this job are given in Sec. 6.3, after we have looked at several specific examples.

Moon predictions are needed for daylight sights, when you can't find the moon with the naked eye because there is insufficient contrast between moon color and sky color. Sun predictions are needed when you are figuring the best time of day for a sun-moon fix or possibly sun-Venus fix (as discussed later in Chapter 6) and for figuring the minimum time you must wait between successive sun sights for a running fix.

Even if you find limited use for the Star Finder in precomputing stars for routine sights, the value of it in identifying unknown stars after the sight (Application I, Chapter 4) is more than enough reason for having one. Furthermore, the Star Finder is the most convenient method for planning and optimizing sun-moon sights. Following examples will compare star precomputation by all possible methods, which should help you choose the method you prefer.

5.1 Precomputing Stars for Routine Sights

This procedure is almost identical to that used for identifying unknown stars.

STEP 1. Predict the UTC of the beginning of twilight—the time you plan to begin taking the sights—using the *Nautical Almanac* and your DR position. Instructions for doing this are in the Almanac; the procedure is explained with examples in Section 3.5. Use the time of civil twilight for evening sights, and the time of nautical twilight for morning sights. These times are listed in LMT and they must be converted to UTC using your DR longitude. Find the GHA of Aries at the UTC of twilight from the *Nautical Almanac*, and then use your DR longitude to find the LHA Aries, exactly as done in the star identification application.

STEP 2. Pick the proper blue template (closest to your DR latitude), mount it on the white disk, and rotate it till the blue arrow points to the proper LHA Aries on the white disk rim scale.

STEP 3. Now make a list of the names, heights, and bearings of all magnitude one stars with heights between about 15° and 75°. Try to estimate the heights to the degree, but precise bearings are not important. Reasons for limiting routine sights to the height range of 15° to 75° and other criteria for choosing optimum stars are discussed in Sec. 5.2.

STEP 4. The next step is to pick 3 of these stars that are roughly evenly spaced around the horizon. If we had the ideal choice, they would be at about the same height and 120° apart in bearing. When making the list, recall that stars at the same bearing will give the same Line of Position (LOP), so you only need to list the brighter one if two lie on the same bearing line. Stars with directly opposite bearings also give the same LOP, but you can't rule out any of these at this stage since you won't know which of the two makes up the best triangle with other available stars.

One way to pick the best three is to plot their azimuth positions on the rim of the compass rose of a plotting sheet. From this picture it is fairly easy to spot the three that are most evenly spaced. In rare cases, if you want the optimum three-star fix, you may have to return to the Star Finder and include a few magnitude two stars. In practice, though, the 20 or so brightest stars (together with available bright planets) will offer sufficient selection for good sights.

STEP 5. At this stage you have two choices. One is to simply consider the job done. Make a list of the heights and bearings of your three stars, wait for twilight, and take the sights. In most cases this should work—that is, your predictions direct from the Star Finder should be accurate enough to get the stars and horizon into view as you predicted them. But to be sure of this, you might save time in the long run by doing a sight reduction of each of the predicted stars from your actual DR position. This extra step removes the uncertainty that comes from using only the closest blue template, not your actual DR latitude, and it also removes errors that come from reading the blue scales. It also serves as a double check that you have come up with favorable stars that are going to be where they are supposed to be.

If you do your sight reductions with a calculator, such as the StarPilot (starpilotllc.com), this extra work takes only a few minutes, and it could make the difference between a successful session of star sights and a frustrating one. It will also pay to take your precomputed list up on deck some time before twilight to see that the stars you chose will be visible from the boat. That is, that the sky and horizon look clear in all directions you care about, and that there will not be any sails blocking a favorite star. Generally one should avoid going forward of the mast with a sextant in rough water.

This procedure of picking optimum stars may seem involved at first, with the azimuth plotting, extra sight reductions, and so forth, but it is worth the time and trouble. Furthermore, you typically only have to do it once or twice, at the beginning of your cruise. Sky cover permitting, you will usually use the same few stars, or star-planet combinations, throughout an ocean crossing. Once you know what these bodies will be from your initial Star Finder work, for the rest of the cruise you can do quick precomputation by direct sight reductions of these bodies from your DR position, without use of the Star Finder.

5.2 Choosing Optimum Stars for Routine Sights

Whenever possible, we choose stars above 15° in height to avoid the uncertainties in atmospheric refraction that affect low sights in unusual atmospheric conditions. Even after the special corrections for low-altitude sights given in Table A4 of the *Nautical Almanac*, sights taken within a few degrees of the horizon could be in error by as much as 5 or 10 miles. On the other hand, if no other sights are available, we should, of course, take whatever sights we can get—it is unlikely that refraction errors, after corrections from Table A4, would be much larger than 10 miles or so, even for sights right on the horizon.

Choosing stars with heights below 75° has two advantages. First, higher sights are somewhat harder to take accurately, and the difficulty increases rapidly as the height approaches 90°. The problem with near-overhead sights is, you can't tell which way to point the sextant when you rock it. The second problem with very high sights is the special plotting procedures required for sights approaching 90°. Standard LOP plotting procedures are not accurate for sights above some 85° or so. See *Celestial Navigation: A Complete Home Study Course*.

Choosing three stars near 120° apart has the important effect of essentially removing any constant errors in the individual sextant sights, such as a misread index correction or a personal bias in aligning the star with the horizon. The center of the triangle of three LOPs, 120° apart, will be an accurate fix, if the only error present is a constant one that affects each sight the same way.

This is also the reason for choosing three stars at about the same height whenever possible, but this criterion for star choice has the lowest priority—meaning in practice it doesn't make much difference. If there were any errors in refraction, they would likely be the same for stars of the same height, so they would cancel out in a three-star fix.

Technical note: sights of three stars 60° apart will yield a triangle of LOPs that appears identical to one obtained from three stars 120° apart. In this case, however, the constant errors do not mathematically cancel when you choose the center of the triangle as your fix. In most practical circumstances, however, this distinction is rather academic, since the errors may well be different for the three sights, with their three different horizons.

Choosing bright stars whenever possible is obvious. They are easier to see while the horizon is still sharp. A typical magnitude one star is about six times brighter than an average magnitude three star. But there are extremes within this average. Vega, for example, is 15 times brighter than magnitude three stars—or even more extreme, Sirius is 75 times brighter than Acamar, though both are navigational stars.

Example 5-1
Precomputation for Star Sights

You are planning for star sights during the morning twilight on July 14, 1982. Your DR position at that time will be 42° 39' N, 57° 12' W. Make a list of stars that might be used and choose the 3 optimum ones.

From the *Nautical Almanac* (see Appendix) the LMT of morning nautical twilight at latitude 40° N (the tabulated latitude closest to your DR-Lat) is 03h 30m. To figure what the UTC will be at that time, first convert your DR-Lon to time using the Arc to Time table in the *Nautical Almanac*. In this case, longitude 57° 12' = 3h 48m 48s, which is about 3h 49m. Thus we expect to be doing our sights at UTC = LMT + DR-Lon(W) = 3h 30m + 3h 49m = 6h 79m = 0719 UTC on July 14th.

Now figure the LHA Aries at UTC 07h 19m. From the *Nautical Almanac* the GHA Aries at 07h 19m is 41° 35', so LHA Aries = GHA Aries − DR-Lon(W) = 41° 35' − (57° 12') + (359° 60') = 343° 83' = 344° 23' = 344.4°. (We write 360° = 359° 60' because we are subtracting 57° from 41°.) Place the 45° N blue template on the north side of the white disk and set the blue arrow to 344.4°. Now make a list of the magnitude one stars with heights between 15° and 75°. You should find the Star Finder results listed in Table 5-1. We will discuss the other data in the table later on.

The relative bearings of these stars are shown in Figure 5-1. It is obvious that Vega, Capella, and Fomalhaut make a near ideal set of stars at about 120° apart. These are the same 3 chosen by Pub. 249 Vol. 1, and we can now see why. Vega is chosen over Deneb because it is lower and much brighter.

After choosing these three stars, you could double-check your predictions by doing sight reductions from your exact DR position. The results you would get are also shown in Table 5-1. In this case it turns out that this extra step was not required. The exact heights are all within 2° or so of the Star Finder predictions, well within sextant view. Nevertheless, it is good practice to make this check, especially if your DR-Lat had been about 40° N, halfway between the 45° and 35° templates. In general, we use the Star Finder to choose the stars, and then the sight reductions (computed or tabulated) to get the actual precomputed values for the sights.

For comparison, Table 5-1 also shows the results you would get if you used Pub. 249 Vol. 1 for this

Table 5-1. Comparison of Three Methods for Precomputing Stars

	Star Finder[a]		Exact[b]		Pub. 249 Vol. 1[c]	
	Hc	Zn	Hc	Zn	Hc	Zn
Capella	28°	052°	26° 35.1'	050.8°	26° 36'	051°*[d]
Aldebaran	16°	082°	15° 10.4'	081.4°	N/A	
Fomalhaut	15°	181°	17° 38.2'	180.2°	17° 16'	180°*
Altair	36°	243°	36° 54.4'	244.5°	36° 59'	244°
Deneb	66°	282°	65° 26.7'	287.8°	N/A	
Vega	42°	287°	41° 36.1'	288.7°	41° 58'	288°*

(a) Using LHA Aries = 344.4° and the 45° N template. (See Appendix for note on improving this reading.)
(b) From sight reductions at the precise DR position (42° 39' N, 57° 12' W).
(c) Using LHA Aries = 344° and Lat = 43° N.
(d) Asterisks mark the optimum 3 stars picked by Pub. 249.

same problem. The asterisks show that the Pub. 249 recommends the same stars we chose. The Pub. 249 results are more accurate so the extra sight reductions are typically not required. Remember, though, that Pub. 249 Vol. 1 can't predict planets. When we want to choose optimum star-planet combinations, the Star Finder plus sight reductions is more accurate and nearly as fast as the tables for this job. In Example 5-6 we shall see that the moon and Venus were also available for sights during this session. If we chose to use either of these, we might pick other stars to complete the triangle.

To show an example of a worst case—a case where we would expect the Star Finder to be the farthest off the exact values—consider the same DR-Lon and UTC, but now at DR-Lat = 40° 10' (halfway between two blue templates). In this case we would still use the 45° template for the Star Finder and get the same Star Finder results given in Table 5-1. The accuracy comparison is shown in Table 5-2.

At this latitude the Star Finder height of Fomalhaut is off by 5°, which means it might well be out of view. We might still find it in calm seas by setting the sextant a few degrees above and below the prediction when it did not show up as expected. But we might not. In smooth or rough water, however, it is best to have it right to begin with. See the Appendix for a simple template adjustment to improve the accuracy of the Star Finder reading based on DR-Lat, and keep in mind that DR-Lat itself can be uncertain.

Publication 249 Vol. 1, on the other hand, still did a good job, because it lists data for every 1° of latitude. The basic conclusion here is, the Star Finder does a pretty good job of star precomputation even in a worst case situation, but when our latitude is not

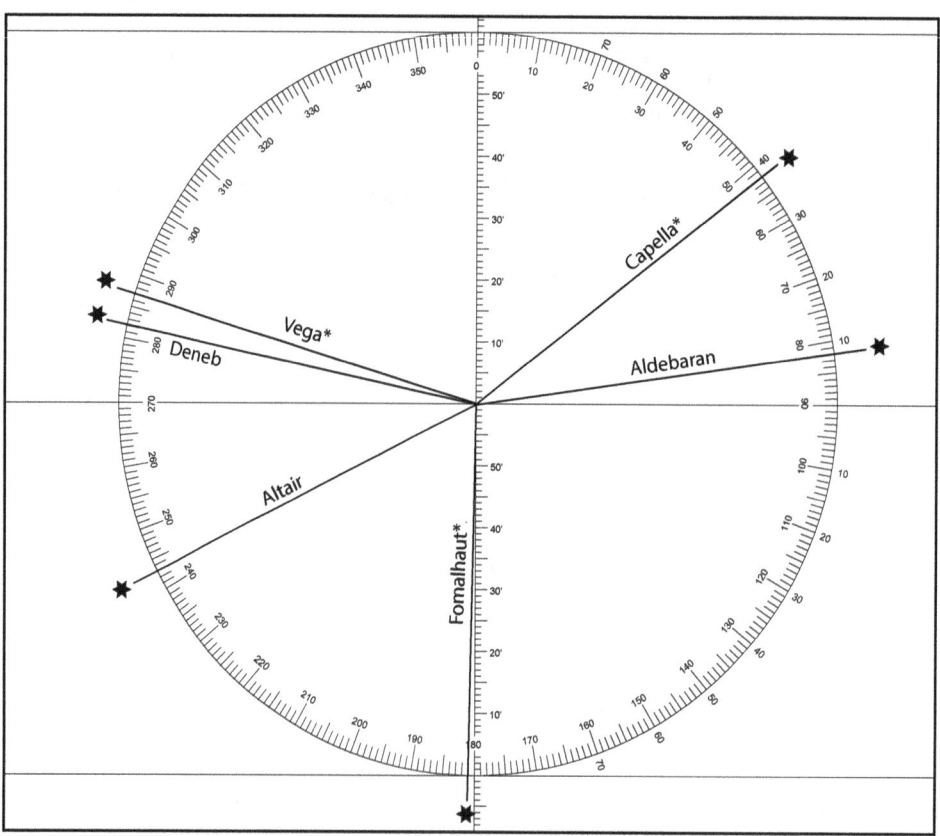

Figure 5-1. *Relative bearings of precomputed stars in Example 5-1. Asterisks mark the 3 stars chosen by Pub 249.*

Table 5-2. Worst Case Comparison of Star Precomputation

	Star Finder[a]		Exact[b]		Pub. 249 Vol. 1[c]		Star Finder with Improved Accuracy[d]	
	Hc	Zn	Hc	Zn	Hc	Zn	Hc	Zn
Capella	28°	052°	25° 00'	050°	24° 41'	050°	24°	049°
Fomalhaut	15°	181°	20° 07'	180°	20° 16'	180°	21°	181°
Vega	42°	287°	40° 46'	291°	40° 57'	291°	40°	292°

(a) Results from 45° template presented in Table 5-1. (b) From sight reductions at DR-Lat 40° 10' N. (c) From Lat 40° N. (d) See Appendix for improved accuracy technique.

near a blue-template value it is best to follow up the Star Finder with a quick sight reduction to get the extra accuracy.

5.3 Precomputing Sun, Moon and Planets

Use of the moon and planets was discussed earlier in Section 3.4. We need to precompute Venus and Jupiter routinely since they are easy and accurate sights that should be taken whenever available. The need for moon precomputation arises mostly when doing daylight sights, either when it is hard to find the moon, or we want to optimize sun-moon fixes. Occasionally it is also helpful to check whether a crescent moon will be high enough to combine with star sights at twilight.

We need sun precomputations for optimizing sun-moon sights, but there is also another use. It is convenient to have the sun plotted on the white disk during the day because we can then tell in a moment what the sun's bearing will be at any time. From this we can figure how long to wait between successive sun lines in order to get a good sun-sun intersection angle for a running fix—the sun's bearing should change by at least 30° or so if we are to get a good fix. The rate that the sun's bearing changes throughout the day depends on several factors and it is not easy to guess this without precomputation.

Precomputing any of these celestial bodies boils down to plotting their proper positions on the white disk. Once that is done, you precompute their heights and bearings the same way you do a star's.

Of these three bodies, the planets are the easiest to plot because their SHAs are listed directly on the daily pages of the *Nautical Almanac*, bottom right hand corner of the "Planets side" of the daily pages, under the star column, as shown in the Appendix. The declinations of the planets are listed each hour, but they don't change rapidly. It doesn't matter what hour you take their declinations from.

STEP 1. To plot a planet on the white disk, look up its Dec and SHA in the Nautical Almanac, and then use the red template to find this position on the white disk as follows.

STEP 2. Put the red template on the white disk, and rotate it until the arrow head on the slotted line points to the value of 360° − SHA. Then within the open slot, use a pencil to mark the planet location at the proper declination. Remember that you don't set the red arrow to SHA on the rim scale of the white disk, but rather to 360° − SHA; and in the Northern Hemisphere (north side of white disk) northern declinations are toward the center of the disk in the solid red lines. In short, to plot a planet on the white disk, you do the planet identification procedure backwards. Once the planet is plotted, look up the LHA of Aries at the time of interest, and precompute its height and bearing with the blue template just as if it were a star printed on the disk.

STEP 3. The sun and moon can be plotted on the white disk in the same manner, once we know their SHAs at a given time. To figure the SHA of the sun or moon, first look up its GHA in the *Nautical Almanac* at the time and date you care about, also record the value of the GHA Aries at that same time. Then figure the SHA from the formula:

SHA (sun or moon) = GHA (sun or moon) − GHA Aries.

If this value turns out to be negative, add 360° to it to find the proper SHA. Declinations of the sun and moon are listed for each hour. The plotting procedure is then the same as for planets. The sun and moon will also always lie within the open slot of the red template. Once plotted, they can be precomputed with the same procedures used for stars and planets.

Example 5-2
Precomputing Sun, Moon, and Planets

In this example we consider a daylight fix using sun and moon, and we also plot and check the location of Venus. Or, put another way, this is simply an exercise in using the Star Finder to answer the question: what are the heights and directions of the sun, moon, and Venus at some specific time.

We consider a mid-morning sight at WT = 09h 50m; our DR position is 35° 10' N, 140° 05' W, on July 14th, 1982. The Zone Description of the watch is +9 hr and we will neglect its small Watch Error for this prediction. First step figure the UTC = WT + ZD = 09h 50m + 9h = 1850 UTC on July 14th. Now from the Almanac we find the GHAs and declinations of the sun, moon, and Venus.

Almanac data at 1850 UTC, interpolated between 1800 and 1900 (see Appendix) are:

	Dec	GHA − GHA Aries + 360° = SHA
Sun	N 21° 38.8'	101° 03.1' − 214° 48.4' + (359° 60') = 246° 14.7'
Moon	N 06° 41.4'	185° 11.5' − 214° 48.4' + (359° 60') = 330° 23.1'
Venus	N 22° 13.2'	132° 26.9 − 214° 48.4' + (359° 60') = 277° 38.5'

Note that the SHA we find for Venus is not exactly the same as the value listed on the July 12-13-14 daily page. Ours, however, is more accurate; the single daily page value of 279° 56.2' = 279.9 (in the bottom-right corner, see Appendix) is for the middle day of the page (July 13) at 00h UTC. Typically we can ignore this difference and take the planet SHA directly from the daily page—the difference will rarely be much larger than the 2° or so we see here.

The values we need to plot the body positions on the white disk are:

	Dec		360° − SHA = Rim Scale
Sun	N 21° 38.8' = N 21.6°	360° − 246.2° = 113° 45.3' = 113.8°
Moon	N 6° 41.4' = N 6.7°	360° − 330.4° = 29° 36.9' = 29.6°
Venus	N 22° 13.2' = N 22.2°	360° − 277.6° = 82° 21.5' = 82.4°

Place the red template on the north side of white disk. Again, either side of red template can face up, but the South side is more convenient since the Dec scale at the slot is easier to read. To plot the sun, set the red arrow to 113.8° and put a dot on the white disk through the slot next to N 21.6° (just below the 22° mark)—north declination means on the solid-circles side of the 0° Dec line. Likewise plot the moon position with the arrow set to 29.6° and Dec about N 7°, and Venus at 82.4° and about N 22°. Now the positions are plotted. Remove the red template and embellish the dots with the conventional symbols (or initials) for these bodies. You can check your plotting relative to the stars in Figure 5-2.

Now we figure the rim scale reading for the white disk that corresponds to WT = 0950 (UTC = 1850). The white disk rim scale = LHA Aries = GHA Aries − DR-Lon(W) = 214° 48.4' − 140° 05' = 74° 43.4' = 74.7'. Put the 35° blue template on the white disk and set the blue arrow to 74.7°. We can now read the heights and bearings of Sun, Moon, and Venus at this time of day. These are compared to exact calculations in Table 5-3.

Differences between the "Exact" and "Star Finder" results represent the author's errors in plotting and reading the scales. The agreement is good, as it should be, since the DR-Lat used for the sight reductions was the same as the blue template latitude; the error in the Venus bearing shows that bearings are sensitive to precise plotting and reading for high altitude sights. But remember, precise bearings are not critical for precomputation. If we have the height right we can find the body by scanning the approximate direction.

Note that once we have the Star Finder set up for 0950 WT, we can read the heights and bearings of these bodies throughout the day. Further Star Finder applications rely on this basic procedure for precomputing heights and bearings of sun, moon, and planets.

5.4 Choosing the Time Between Sun Lines for a Running Fix

We have discussed this topic briefly in earlier sections and will now illustrate it in detail with an example. This is a clever and useful application of the

Chapter 5: Application II

Table 5-3. Precomputed Sun, Moon, and Venus

	Star Finder[a]		Exact[b]	
	Hc	Zn	Hc	Zn
Sun	53°	102°	53° 25'	101°
Moon	40°	248°	39° 52'	247°
Venus	76°	155°	75° 35'	150°

(a) As I read my Star Finder from my plotting. Done carefully, but without "fudging" once I calculated the exact values. (b) Exact values calculated by sight reduction from 35° 00' N, 140 05' W at 1850 UTC, July 14th, 1982.

Star Finder that should prove valuable on any cruise. As we shall see below, it is especially valuable when the midday sun is high.

The goal is to find the time of day when we can take 2 successive sun lines that will intersect at an angle of at least 30° with the minimum time between sights. If the lines intersect at any narrower an angle, the fix will be a weak one—meaning small errors in sight or plotting result in a large error in the fix. The general topic of LOP intersection angles and celestial bearings is discussed in more detail in Chapter 6.

Before starting we should note that when the height of the midday sun is less than about 45° high you do not need the Star Finder or any other aid for this problem. You can use a simple rule of thumb for picking the time between sun sights. Just assume the sun's bearing changes by 15° per hour throughout the day. In practice it will be somewhat less than this in the mornings and afternoons, and somewhat more near midday. But for purposes of timing running fixes this is a reasonable approximation at any time of day when the sun at noon is low. In this case—noon sun less than halfway up the sky—you need to wait 2 to 3 hours between sun lines for a running fix regardless of the time of day, and you will be guaranteed that the two sun lines will intersect by at least 30°. The noon height of the sun depends on your latitude and the sun's declination (the time of year). One could make up formulas using latitude

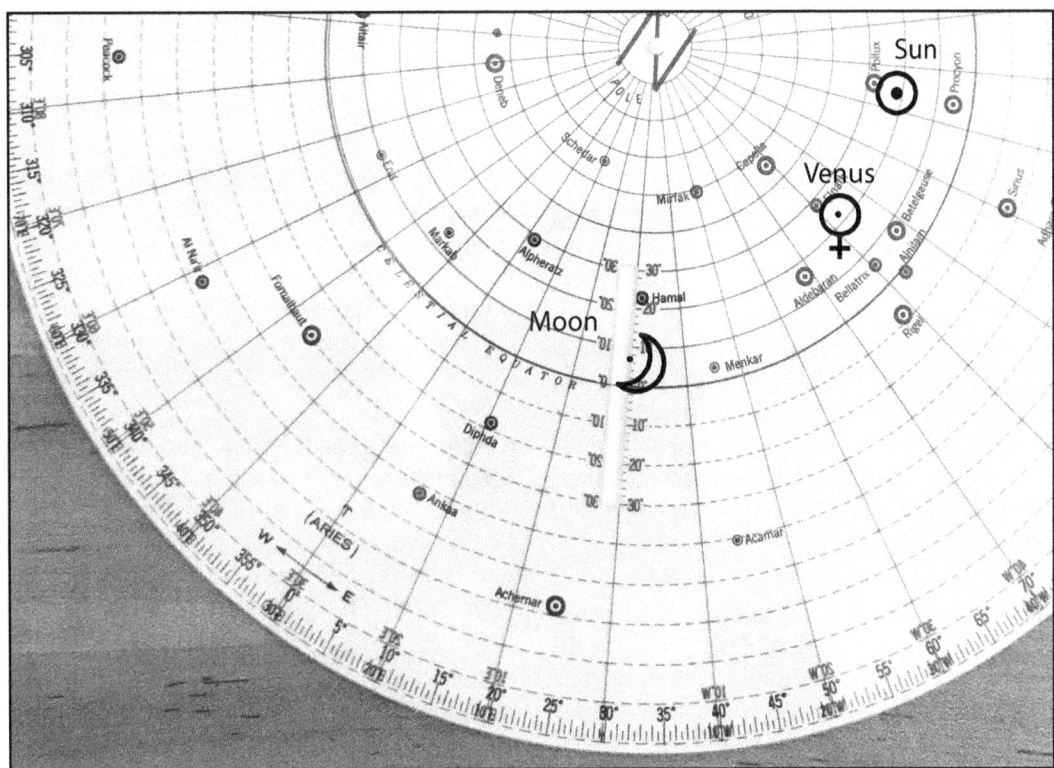

Figure 5-2. *Sun, moon, and Venus plotted on the white disk for use in Example 5-2.*

Example 5-3
Picking the Time Between
Sun Sights for a Running Fix.

In this example we will use the sun position already plotted on the white disk in Example 5-2. The date is July 14, 1982. We first note that the position of the sun (or planets) on the white disk does not at all depend on our latitude and longitude. Sun, moon, and planet positions on the disk depend only on their latitude and longitude (Dec and SHA), which in turn depends only on the date and time. Also, for the sun it doesn't matter much what time of day we use to plot its position on the disk since it only moves 1° per day across the disk, through the stars.

To illustrate this, in Example 5-2 we used a UTC of 1850 and got a Dec = N 21.6° and an SHA = 246.2°, which gave us a rim scale location = 360 − 246.2 = 113.8°. Now let us plot the sun's position on this same date using a time of 0100 UTC. From the Almanac we find at 0100 UTC on July 14, 1982 that the sun's Dec = N 21° 45.4' = N 21.8° and the sun's GHA = 193° 34.3'. The GHA Aries at 0100 = 306° 34.4'. From these we find the sun's SHA = GHA sun − GHA Aries = 193° 34.3' − 306° 34.4' (+ 359° 60') = 245° 59.9' = 246.0°. This gives a rim scale of 360 − SHA = 360 − 246.0 = 114.0°. From this we can see that from 0100 to 1850 the sun's coordinates on the white disk only changed by a few tenths of a degree. The conclusion is, when plotting the sun you can use whatever time of day is most convenient for you. But this is not the case for the moon and planets, especially the moon which can move some 12° or 13° across the disk during one day.

So now we start. With the sun plotted on the white disk at Dec about N 22° and rim scale about 114°, we will look at the motion of the sun throughout the day from several latitudes, and from this judge how long we must wait between sun lines for a good running fix. The answer will depend on our latitude.

Example 5-3 (a)

July 14, 1982; DR-Lat = 48° N, DR-Lon = 155° W. The ZD of our navigation watch is +10 hr. Checking the Almanac (see Appendix) we find sunrise (at Lat = 50° N, the closest one listed) is at 0405 LMT and sunset is at 2006 LMT. Our DR-Lon of 155° W = + 10h 20m when converted to time with the Arc to Time Table. So sunrise is at UTC = 0405 LMT + 1020 = 1425 UTC. For convenience, we start our look at the sun beginning at UTC 1500. At 1500 UTC the GHA of Aries is 157° 08.9', so the LHA Aries = 157° 08.9' − 155° 00' = 002° 08.9' = 2.2°. Watch time = UTC − ZD = 1500 UTC − 1000 = 0500 WT. So to set up the time scale on the white disk, mark the 2.2° position with WT = 0500, and mark every 15° from there with the next hour: 17.2° = 0600, etc.

Place the 45° N template (the closest one to our DR-Lat of 48° N) on the white disk and rotate it through the day. As you do this make a list of the heights and bearings of the sun at each hour. You should get the approximate values given in Table 5-4. The table gives values for every 30 minutes to show in more detail how the sun moves. The table also shows these same results for two other latitudes (15° N and 45° S, both also at Lon 155° W), which we will compare shortly.

Referring to your list or Table 5-4, at latitude 45° N on July 14th we might get our first good sun line at about 0700 WT, since that is when the sun first reaches a height of 20° or so. For the next few hours after that the sun's bearing changes at a rate of about 10° per hour. If we want an LOP intersection angle of at least 30° we must wait about 3 hours, or until about 1000 WT.

Looking at a day's navigation in this case, we might take a star fix at about 0400 WT and then finish a running fix from the sun at about 1000 WT. We could then get another running fix between 1400 and 1600 WT and finish with evening star sights at about 2100 WT. This makes up a good day's navigation.

On a typical day, however, star fixes in the morning and evening plus just one running sun fix during the day might well be enough navigation. But if the

Table 5-4. Sun's Height (He) and Bearing (Zn) on July 14, 1982[a]

		Lat 45° N			Lat 15° N			Lat 45° S		
GMT	WT[b]	Hc	Zn	dZn[c]	Hc	Zn	dZn	Hc	Zn	dZn
1500	0500	1	060							
1530	0530	6	065							
1600	0600	11	070	10	0	067				
1630	0630	16	075		6	069				
1700	0700	21	080	10	13	071	4			
1730	0730	26	085		20	072				
1800	0800	32	090	10	27	073	2	0	058	
1830	0830	37	096		34	074		4	053	
1900	0900	42	102	12	41	074	1	9	047	11
1930	0930	47	108		48	074		12	041	
2000	1000	52	116	14	55	074	0	16	035	12
2030	1030	57	125		62	072		18	028	
2100	1100	61	135	19	69	069	5	21	021	14
2130	1130	64	149		75	061		22	014	
2200	1200	66	165	30	81	042	27	23	007	14
2230	1230	67	182		83	352		23	359	
2300	1300	66	200	35	79	311	91	23	351	16
2330	1330	63	215		73	296		22	344	
0000	1400	60	227	27	67	290	21	20	337	14
0030	1430	55	238		60	287		18	330	
0100	1500	51	246	19	53	286	4	15	323	14
0130	1530	46	254		46	286		11	217	
0200	1600	41	260	14	39	286	0	7	311	12
0230	1630	35	266		32	286		3	305	
0300	1700	30	271	11	25	287	1			
0330	1730	25	277		18	288				
0400	1800	20	281	10	11	290	3			
0430	1830	14	286		5	291				
0500	1900	9	291	10						
0530	1930	5	296							

(a) DR-Lon = 155° 00' W; all values in degrees, all bearings are true.
(b) ZD = +10 hr.
(c) dZn = change in Zn during the previous hour, in degrees.

skies are cloudy or questionable, and we suspect we might miss the evening star sights, then the extra afternoon running fix would be good insurance.

Another option in this case is to use this Star Finder data to choose one sun line and then do an LAN latitude sight. LAN occurs at about 1230 WT, and near this time (at this latitude) the sun's bearing changes at about 30° per hour before and after.

So we could get a running fix from one sun line and an LAN latitude near midday, and it would only take about an hour or so. If we intend only one daytime fix this would be an efficient option.

We could also do two sun lines near midday about an hour apart and still get a midday fix without resorting to the LAN sight. The LAN sight has a distinct disadvantage of requiring us to do something at a specific time – we must catch and measure the maximum height of the sun. Whereas the precise times of the two sun lines near midday are not critical.

The general goal in picking times for running fixes is to have the time between sights be as short as possible to minimize the effects of DR errors. The accuracy of your running fix is limited by the accuracy of your DR between the two sun lines. If you figure you ran 20 miles between sights, but your log or knotmeter was wrong, or you had significant currents present, and you actually ran 23 miles, then your running fix will be in error by roughly 3 miles.

Example 5-3 (b)

Same date, same watch, same longitude, but now viewing the sun from latitude 15° N. This latitude puts you just south of the sun, with the noon sun passing high overhead at a maximum height of some 83°. This height is not too high for an LAN sight, but it will take some practice to get accurate sights in this case if the water is rough. Table 5-4 shows that a running fix from the sun is not possible during the morning or during the afternoon. The bearing of the sun changes little all morning, then it pops overhead, changing directions rapidly during the hour before and after noon, and then changes bearings very slowly again in the afternoon.

If you were confident with your high sun sights, you could do a LAN-sun line running fix during the hour before or after noon. The other option is to take a sun line when the sun is low enough for an accurate sight, at say 1130 WT, and then wait an hour past LAN, till 1330 when the sun is back down again, for a 2-hour running fix with manageable sextant heights. If you have trouble picturing how a large bearing change of the sun (061 to 296, in this case) will translate into LOP intersection angles, the topic is covered more specifically in Chapter 6.

Generally speaking, you will have this way around the high sight problem—running fix from AM to PM—except when you happen to be near the equator within a couple weeks of either equinox. Work this latter problem out with the Star Finder to see that the sun then bears nearly due east all morning and then due west all afternoon. It is just the most exaggerated case of the example just given.

Example 5-3 (c)

Again, same date; same watch; same longitude; but now viewing the sun from 45° S. To check Table 5-4 with your Star Finder you must plot the sun on the south side of the white disk, and then turn over the 45° template so the 45° S side faces up when placed on the south side of the white disk.

This example demonstrates the rule of thumb mentioned above. The noon sun height is less than 45° so its bearing changes by about 15° per hour throughout the day. As soon as the sun is high enough for a sight you must wait just over 2 hours for a bearing change of 30°. Note that for these low suns, there is no significant advantage in time by doing your running fix near midday—though in this extreme case you don't have much of a choice since the day is so short.

5.5 Picking Optimum Star-Planet Combinations

Usually both stars and planets are available for twilight sights. Publication 249 Vol. 1 can predict the optimum three-star fix but it does not take into account the location of the planets. Often the best fix available is not from stars alone, but from two stars and a planet. In the following examples we compare the three-star fix recommended by Pub. 249 with our choice for an optimum fix using both stars and planets. To do this, we predict the heights and bearings of the stars and planets simultaneously using the Star Finder.

Publication 249 chooses its stars by putting upper and lower limits on the heights and then picking the three that are the nearest to 120° apart, using brighter stars over dimmer ones if they are equally well positioned. This choice is made because this triangle yields the most accurate fix—providing everything else is equal. Venus and Jupiter sights, however, are not equal to star sights; they are usually easier and more accurate because they can be taken while the horizon is sharper. Our task then, is to see if we might replace one of the suggested stars with a

Chapter 5: Application II

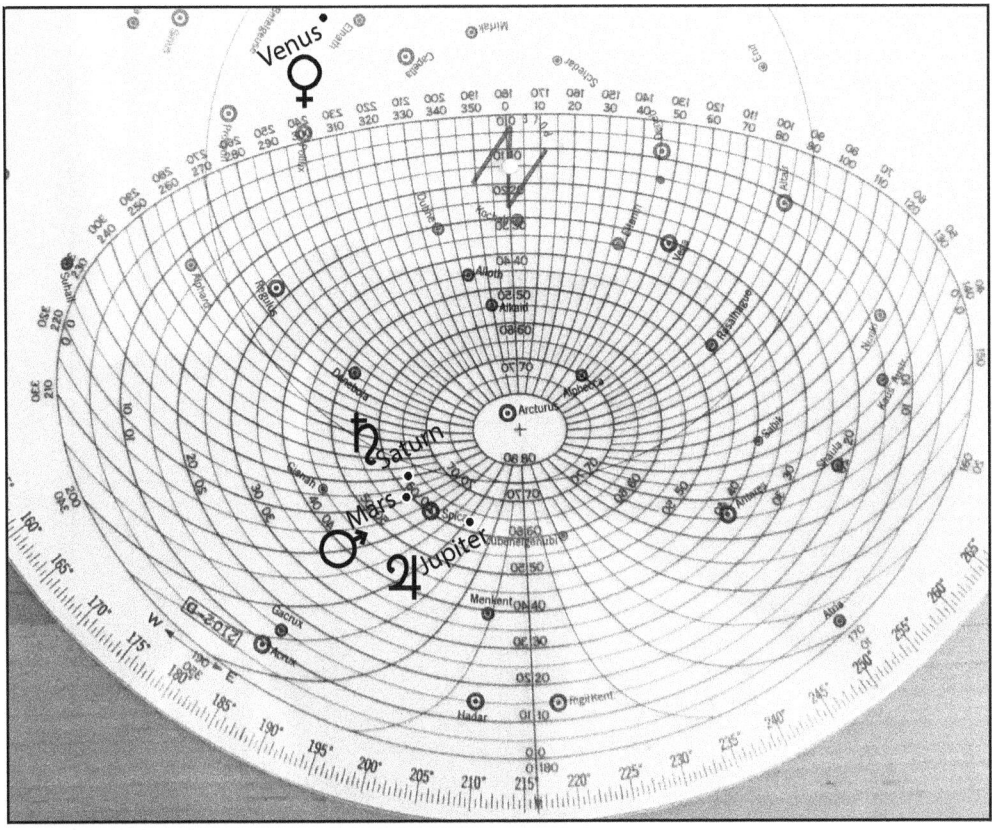

Figure 5-3. *Planets plotted on the white disk for use in Example 5-4. The 15° N template is in place, set to LHA Aries = 216.5°. Compare the height and bearing readings here, or on your Star Finder, to those listed in Table 5-5. Venus is below the horizon; Mars bears about 222° at a height of about 59°; and so forth.*

Data for Example 5-4

Almanac and Star Finder data for the planets on July 14, 1982 (see appendix):
(SHA data from the daily pages; Dec from 1300 UTC)

Planet	SHA	Rim Scale	Declination
Venus	279° 56.2' = 279.9°	360 − 279.9 = 081.1°	N 22° 11.5' = N 22.2°
Mars	163° 04.1' = 163.1°	360 − 163.1 = 196.9°	S 08° 0.6' = S 8.0°
Jupiter	150° 55.5' = 150.9°	360 − 150.9 = 209.1°	S 10° 40.4' = S 10.7°
Saturn	164° 18.5' = 164.3°	360 − 164.3 = 195.7°	S 04° 2.0' = S 4.0°

GHA Aries at 1249 UTC = 124° 18.5'

planet, or to choose the best stars to combine with the planet we want to use, if at all possible. From a practical point of view, using a bright planet when possible will yield a more accurate fix, even if we must sacrifice somewhat on the optimum positioning of the bodies.

Example 5-4
Precomputing Star-Planet Combinations

In this example we want to choose the best combination of stars and planets, and possibly the moon, for evening twilight sights on July 14th, 1982. Our DR position at twilight will be 14° 08' N, 92° 13' E.

The Star Finder Book

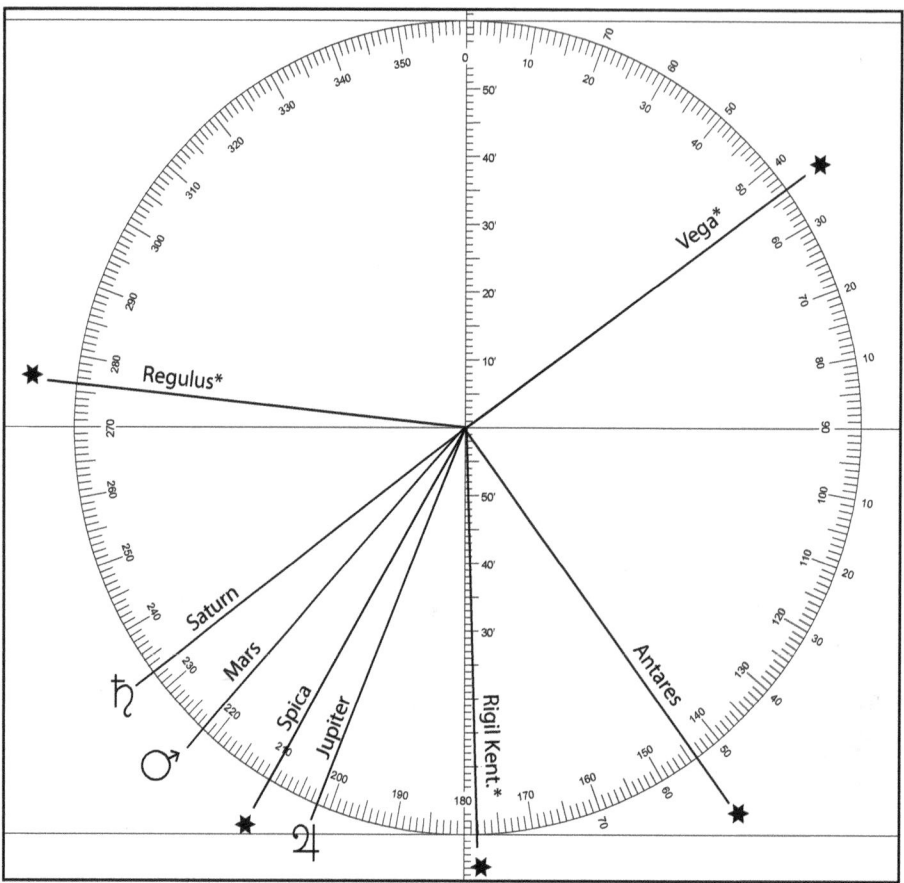

Figure 5-4. *Relative bearings of stars and planets used in Example 5-4. Asterisks mark the 3 stars chosen by Pub 249.*

Table 5-5. Precomputation of Star-Planet Combinations[a]

		Star Finder[b]		*Exact*[c]	
Body	**Mag**	**He**	**Zn**	**He**	**Zn**
Venus	−3.3	—	—	— —	—
Mars	+0.6	59°	222°	58° 33'	222°
Jupiter	−1.7	62°	198°	62° 26'	202°
Saturn	+1.0	62°	230°	59° 54'	233°
Spica	+1.2	60°	212°	60° 02'	210°
Antares	+1.2	40°	145°	38° 44'	144°
Vega	+0.1	30°	053°	30° 11'	054°*
Regulus	+1.3	27°	272°	27° 49'	276°*
Rigil Kent.	+0.1	14°	178°	14° 10'	178°*

(a) On July 14th, 1982, viewed from 15° N, 92° 13' E, at evening twilight, 1249 UTC. Differences between Exact and Star Finder values represent the author's errors in plotting or reading the Star Finder
(b) Read directly from a Star Finder: 15° N template, LHA Aries = 216.5.
(c) Planet Values from sight reductions at 1300 UTC; Stars from Pub 249 Vol. 1 (Epoch 1985), LHA Aries = 216°, Lat 15° N. Stars marked with an * are the 3 recommended by the tables.

Note this is an east longitude example. We first concentrate on picking the best bodies to use; later we figure the actual precomputed heights and bearings for the sights.

We first figure the time of the sights. The Almanac lists the time of evening civil twilight at 10° N as 1848 LMT and at 20° N as 1907 LMT. Our latitude of 14° N is about halfway between these two, so we use the time halfway between: (1848 + 1907)/2 = (1848 + 1867)/2 = 1858 LMT. Our DR-Lon of 92° 13' = 06h 09m, so the UTC = LMT – DR-Lon(E) = 1858 – 0609 = 1249 UTC. Note the minus sign here in figuring UTC from eastern longitudes.

We must use the proper UTC when figuring the heights and bearings, but to plot the planets on the white disk we can simplify the Almanac use by plotting them for the nearest whole hour—1300 UTC in this case. Or, for that matter, we could use any UTC on the right date and we will still get their positions on the white disk accurately enough for this application. The SHAs of the planets are given on the daily pages, and we will use these, rather then figure them at the precise time of the sights.

Use the data given to plot the planet positions on the white disk, as shown in Figure 5-3. You can also refer back to Figure 3-4 to see how these positions appear on the rectangular star maps in the *Nautical Almanac*.

Now figure the rim scale reading that corresponds to UTC = 1249. When used as a time scale, the rim scale = LHA Aries = GHA Aries + DR-Lon(E) = 124° 18.5' + 92° 13.0' = 216° 31.5' = 216.5°. So the 216.5° position on the rim scale corresponds to UTC = 1249—the approximate time we plan to do the evening twilight sights. Place the 15° N blue template on the white disk, set the arrow to the 216.5 position and then make a list of the heights and bearings of the planets and bright stars. Take only the magnitude one stars that are above about 20° and below about 75°. You should get the approximate values given in Table 5-5.

You will find that Venus is below the horizon at this time, and the bright star Arcturus is nearly overhead, too high to use. Nevertheless, we have a large number of bright bodies to choose from and will certainly come up with a good triangle of convenient sights. We include Rigil Kentaurus in the Star Finder list because it was one of the three stars that Pub. 249 recommended.

The moon is not included because it is below the horizon at twilight. You would discover this if you plotted its position on the white disk, but this is not necessary. The Almanac tells us that moonset on July 14th is at about 1230 LMT, which is some 6 hours before twilight (1849 LMT), and it does not rise again till about midnight LMT. So it is clear we need not bother looking for it. You can always make this check on the moon, and then only plot it and consider its value if your sight time is between moonrise and moonset.

The data of Table 5-5 is easier to interpret if we plot it on a Universal Plotting Sheet as shown in Figure 5-4. From the figure we see why Pub. 249 picked the 3 stars it did—they form the best triangle of 3 stars, as close as possible to 120° apart. To get this triangle, however, they had to use Rigil Kent., which is very bright, but at this time is near the limit on height at only 14° high. My own preference in this case would have been to use Antares instead. The triangle is about the same and Antares is higher. Also, even through Rigil Kent. is brighter, Antares is so red that it is just as prominent as Rigil Kentaurus.

Another factor that the Pub. 249 list does not help us with is the presence of another bright star near Rigil Kentaurus, and that is Hadar. The Star Finder would tell us that Hadar (mag. +0.9) is at the same height as Rigil Kentaurus and just 5 or 6° to the west of it. These two stars are the Pointers to the Southern Cross. You can see from the Star Finder that the Cross is above the horizon, directly to the west of the Pointers, but it is too low at this time for routine use in navigation. Since the Pointers are at the same height, and so close together, the usual technique of setting the sextant to the right height and scanning the horizon will not guarantee that you get the right star in this case.

In short, this is a clear-cut case where the Star Finder is more informative than Pub. 249 Vol. 1 for star precomputation. From a practical point of view, Antares is a better choice than Rigil Kentaurus during this sight session. We can learn this from the

Star Finder, but have no way to discover such things from the predictions of Pub. 249 alone.

Now back to the task at hand, picking the best star-planet combination. Since you will typically get more accurate sights using Jupiter in bright twilight with a sharp horizon, my preference is to use it (or Venus) whenever possible. My choice would be to go for Jupiter, Antares, and Regulus; then if I missed Jupiter for some reason (like clouds), take Vega as its replacement.

Example 5-5
More on Star-Planet Combinations

As another example, we will use the same date and DR position, but now look at morning twilight sights. In this case we will get to use Venus which was below the horizon during evening twilight. We can learn this ahead of time—that Venus is a "morning star" during this season—from the Planet Notes at the beginning of the *Nautical Almanac*.

The date is July 14, 1982; the DR position is 15° 00' N, 92° 13' E. Sights will begin at morning nautical twilight, which occurs at about 0456 LMT. Converting to UTC, UTC = LMT − DR-Lon(E) = 0456 − 0609 = −0113. This is 1 hr and 13 min behind 0000 on July 14th, which corresponds to a time of 2247 UTC on July 13th. From the Almanac, the GHA of Aries at 2247 on July 13th is 273° 14.0', so the LHA Aries = 273° 14.0' + 92° 13.0' = 365° 27.0' = 5° 27.0' = 5.5°. Hence the blue arrow on the 15° N template should point to 5.5° when we read off the locations of the stars and planets for our morning twilight sights. The planet positions on the white disk are the same as plotted in Example 5-4. You should get the values given in Table 5-6.

These body positions are plotted in Figure 5-5. In this case it is easy to see that we can replace Aldebaran with Venus and still maintain a good triangle with the other two stars recommended. In both this and the last example, we have ended up simply replacing one of the recommended three stars with a planet. In other circumstances, however, it may be that you must choose two different stars to make up the best star-planet combination. This is most likely when you have many bright stars above the horizon at the same time, and this, in turn, is most likely when the Orion region of the stars is high in the sky at sight time.

Example 5-6
Precomputing Moon, Planet, Star Combinations

In Example 5-1 we picked the optimum three stars for a morning twilight fix at 0330 LMT which corresponded to 0719 UTC from a DR position of 42° 39' N, 57° 12' W. Precomputed stars were listed in Table 5-1. And their relative bearings were plotted in Figure 5-1 in order to choose the best three. We now check to see if the moon or Venus might also be used during this sight session. And if so, does this change our choice of stars?

The LHA of Aries at 0719 UTC is 343.9° as found in Example 5-1. From the Almanac, the locations of the moon and Venus at 0719 UTC on July 14th, 1982 are:

Moon GHA = 18° 01.3'   Dec = N 04° 14.4'

Venus GHA = 319° 50.6'   Dec = N 22° 9.7'

Use these positions to plot the moon and Venus on the white disk, set the 45° N template to LHA Aries = 343.9° and you should find the heights and bearings shown in Table 5-7.

Figure 5-6 compares these moon and Venus bearings with the star bearings at this same time. Only the later Venus bearing is shown because it is too low at the beginning of twilight. The star bearings are the same ones plotted in Figure 5-1. We now have a picture of all bright objects in the sky at this time, and it is easy to see we have several options besides the optimum stars.

Venus is too low at the beginning of twilight, but that does not matter. Our goal, in the first place, is to save it until the brightest part of twilight when the horizon is sharp. We have a good case here when we can take Venus sights after all stars have faded, just before sunrise.

Looking at Figure 5-6, we could take the recommended three stars during early twilight, or, if we sacrifice positioning for easier sights, we could take Fomalhaut, Capella, and a later moon sight. Or per-

Chapter 5: Application II

Table 5-6. More on Star-Planet Combinations[a]

		Star Finder[b]		*Exact*[c]	
Body	**Mag**	**He**	**Zn**	**He**	**Zn**
Venus	−3.3	19°	071°	18° 27'	071
Capella	+0.2	22°	046°	21° 52'	046°
Alderbaran	−1.1	29°	079°	28° 52'	079°*
Achernar	+0.6	16°	171°	15° 56'	169°
Fomalhaut	+1.3	41°	204°	41° 00'	201°*
Altair	+0.9	23°	273°	23° 54'	273°
Deneb	+1.3	35°	315°	35° 11'	315°*

(a) On July 14th, 1982, viewed from 15°N, 92° 13' E, at morning twilight, 2247, July 13th UTC. Differences between Exact and Star Finder Values represents the author's errors in plotting or reading the Star Finder.
(b) Read directly from a Star Finder: 15° N template, LHA Aries = 5.5.
(c) Venus values from a sight reduction at 2247 UTC; Stars from Pub. 249 Vol. 1 (Epoch 1985), LHA Aries = 5°, Lat = 15° N. Stars marked with an * are the three recommended by the tables.

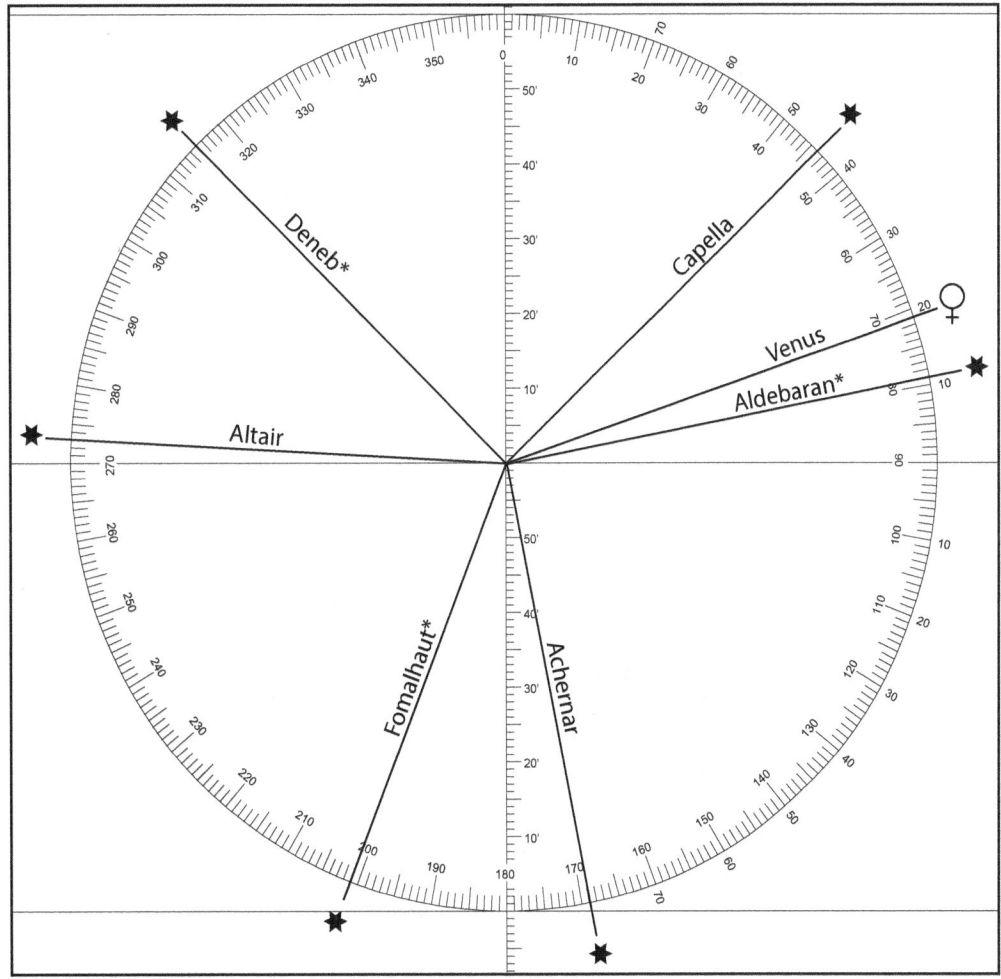

Figure 5-5. *Relative bearings of stars and planets used in Example 5-5. Asterisks mark the 3 stars chosen by Pub. 249.*

The Star Finder Book

haps as good as any, Vega sights toward the end of twilight, and then later moon and Venus sights. This would use the three brightest objects available, with two of them having essentially daylight horizons.

In practice, I would probably take them all—the three best stars (Vega, Capella, and Fomalhaut) throughout twilight, and then the moon and Venus also, after the stars fade. The key here is insurance. Take the star sights while you can and save the brighter objects for a better horizon. If you have limited time, after you have, say, three sights of each star, the star to take the most sights of (an extra two or so) would be Vega. I would not, for example, take the moon sights at all until the star triangle was secure.

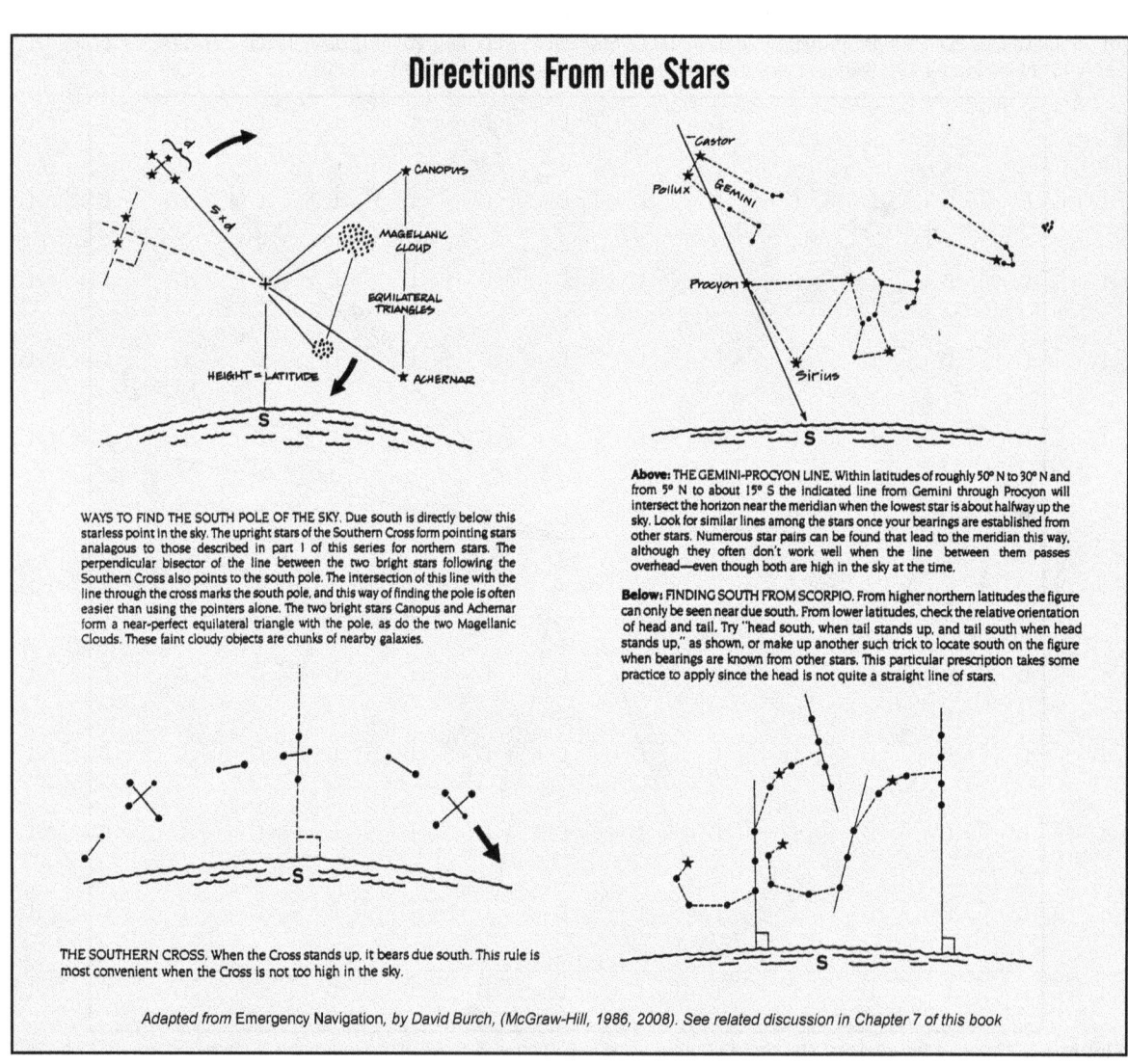

Adapted from Emergency Navigation, by David Burch, (McGraw-Hill, 1986, 2008). See related discussion in Chapter 7 of this book

Chapter 5: Application II

Table 5-7. Data for Example 5-6, 7/14/1982, DR 42°39' N, 57°12' W

	LMT	GMT	Moon		Venus	
			Hc	Zn	Hc	Zn
Start Nautical Twilight	0330	0719	36° 30'	128°	10° 13'	069°
End Nautical Twilight	0410					
Just Before Sunrise	0430	0819	43° 52'	145°	20° 23'	079°
Sunrise	0442					

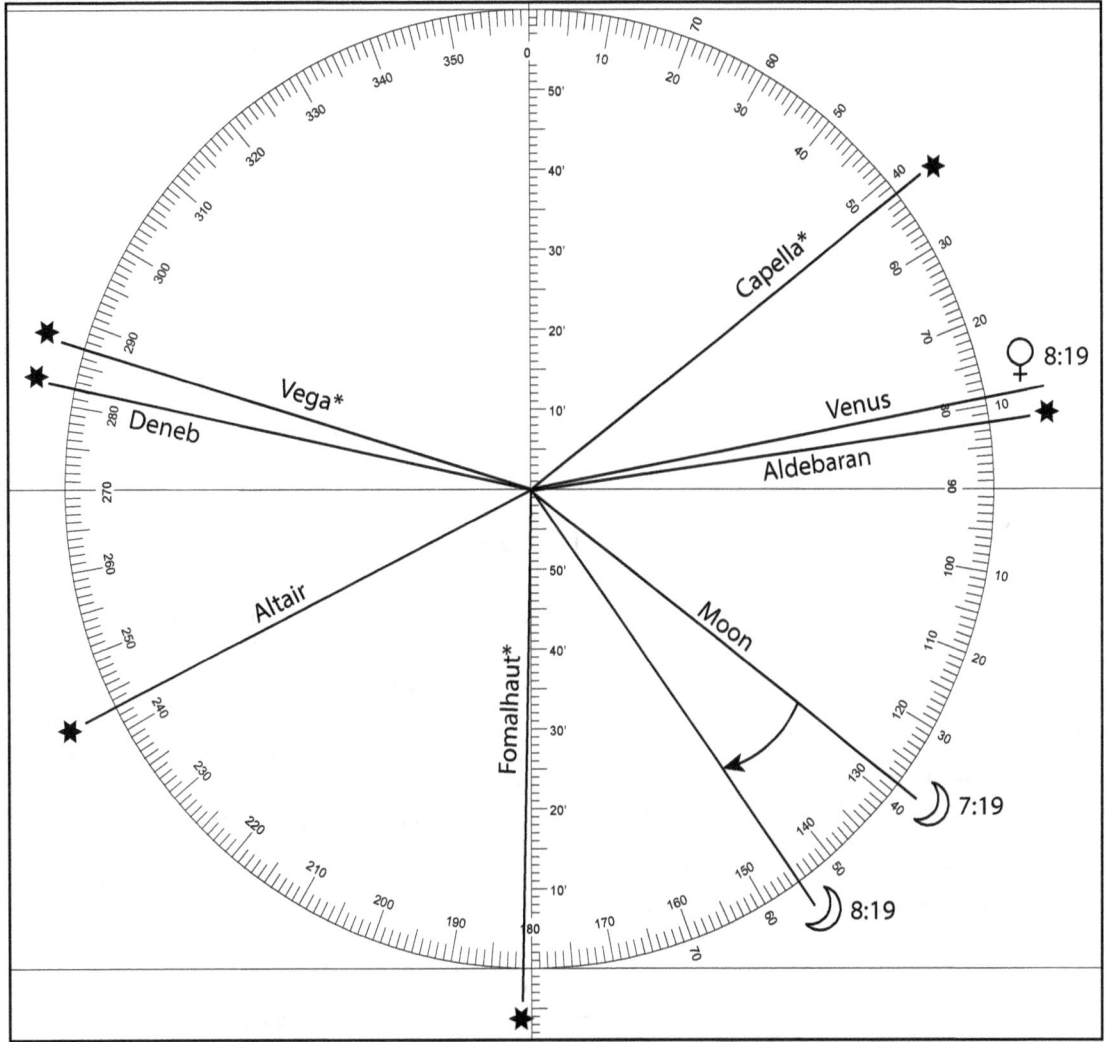

Figure 5-6. *Relative bearings of stars, moon, and Venus used in Example 5-6. Star positions are from Example 5-1. Asterisks mark the 3 stars chosen by Pub. 249, that were made without consideration of the moon and Venus positions. With these bodies available for sights we have other options for a good triangle of sights.*

Chapter 6.
Application III:
Sun-Moon and Sun-Venus Sights

For about 1 to 2 weeks of each month (see Sec. 3.4), the moon provides an excellent two-body fix with the sun throughout much of the day. Beginning navigators often fail to take full advantage of this for typically two reasons: One, they are at first uncomfortable with the sight reduction of moon sights, but this goes away quickly with practice. Two, they are not always aware that the moon is really out there, because a casual view around the sky will not find it—especially if it happens to be temporarily obscured during this search, or it happens to be in a faint phase, or its color is similar to the sky color.

Besides the moon, the planet Venus is the only other celestial body that can sometimes be seen through a sextant while the sun is up. On a few of these occasions, Venus provides a unique opportunity for a two-body celestial fix with the sun. We should admit from the beginning, though, that daylight sights of Venus will rest fairly deep in the navigator's bag of tricks.

The task at hand in this application is to predict when sun fixes with the moon or Venus might be feasible, and, even more specifically, to predict the best times of day to try for sun-moon or sun-Venus sights to get an instantaneous fix. This fix, when feasible, will typically be more accurate than a standard running fix from the sun taken over a period of several hours.

This application is a specialized use of the Star Finder. Its role in practical navigation is discussed further at the end of this section. But despite its sometimes restricted use, this application is an excellent exercise in the use of the Star Finder. Besides the obvious need for precomputation to find Venus during daylight, the Star Finder can also be used to predict—before we do the sights—the intersection angle between the Lines of Position (LOPS) we would get from the sun and moon or sun and Venus. This is vital information because the angle between two plotted LOPS affects the accuracy of the fix. If the angle is narrow (less than 30° or so), then any small error in the sights, analysis, or plotting will result in an enhanced error in the fix.

The optimum crossing angle for two LOPs is 90°, but the problem of enhanced errors reduces rapidly as the angle increases above some 30°. For a rule of thumb, we can consider 30° as the minimum LOP crossing angle for routine fixes—smaller angles can be used in an emergency, but special care and Procedures are required for tolerable accuracy.

The other factor to consider is the heights of the objects. As discussed earlier, optimum heights are between some 15° and 75°—although sights down to 5° or up to 85° are essentially as good in many cases.

As the day progresses, all celestial bodies rise, cross the meridian, then set. But not only their heights and bearings change as they move across the sky, their relative bearings (the angles between them) also change. And the way these relative bearings change is not at all obvious; it depends on too many factors.

In any event, if we are to optimize these sights we need to know the relative bearings of the objects throughout the day, because their LOPs will intersect at this same angle, the difference in their bearings. This comes about because the azimuths (Zn) we get from sight reductions are very nearly the same as the true bearings to the bodies sighted. And since the LOPs are always perpendicular to the azimuths, the angle between the LOPs is the same as the angle between the azimuths. For optimum sights, we need to find the time of day when the heights are right, and the objects are at least 30° apart... but not more than 150° apart, since objects with exactly opposite bearings give the same LOP.

These daylight sights can be especially valuable when sailing in latitudes near the declination of the sun because daytime navigation by sun alone is difficult then: Local Apparent Noon sights for latitude are difficult when the sun passes near overhead, and successive morning or afternoon sun lines intersect at narrow angles. Accurate running fixes from the sun in these circumstances are difficult to do. In short, whenever or wherever you sail under the sun, you will be happy to have a daytime moon for navigation between twilight star fixes.

6.1 Figuring Optimum Sight Times

Note: we cover the Venus sights here along with the moon sights since the procedures are very similar, but the moon sights will be used much more often than the Venus sights.

STEP 1. First check Table 3-3 (in Sec. 3.4) to get a first guess at whether and when the moon will be useful. Then turn to the *Nautical Almanac* daily pages for the day in question, and list the rising, meridian passage (mer pass), and setting times for sun, moon, and Venus. For Venus only the meridian passage time is listed, but its rising and setting times can be estimated by comparing its mer pass time with the suns. Assume the time difference in mer pass also applies to rising and setting: if Venus crosses the meridian 1 hour before the sun, assume it will rise and set 1 hour before the sun. This won't be exact, but we don't need to know this precisely.

Convert all these times to UTC using your DR-Lon, and from these times figure the UTC time interval when the bodies you want will be above the horizon at the same time. Then estimate the time, to the nearest hour, that is about midway through this interval.

STEP 2. Use this midway UTC to look up the Dec and GHA of these bodies, and also record the GHA of Aries at that UTC. Figure the SHAs and then rim scale values for each body and plot them on the white disk, as described in Chapter 5. Also label the rim scale with UTC, using LHA Aries on the rim scale equal to the midway UTC you chose.

STEP 3. Now place the appropriate blue template on the white disk and rotate, it to check that the sun rises at the right time: the blue arrow should point to the UTC of sunrise (or very near it) when the sun first enters the blue scale diagram, at Hc = 0°. Also check sunset, and do both checks with the moon and Venus.

Likewise check that sun, moon, and Venus cross the meridian at their predicted times as you rotate the blue template. When each of these crosses the blue line with the arrow on it (bearing either 180° or 360°) the blue arrow should point to the UTC of mer pass that you figured from the Almanac time of their meridian passage. Note that the moon or Venus may not cross the meridian during the day—meaning when the sun is above the horizon, within the blue scale—but that doesn't matter; you can still make this check. This is, after all, one of the reasons we are doing this, to see when they are where. This first step is just a double check that the body positions are plotted correctly on the white disk.

STEP 4. Now rotate the blue template from sunrise to sunset and record the respective heights and directions of sun, moon, and Venus at one hour intervals throughout the day – or throughout the time period suggested by Table 3-3. Once you have this list, make a new list of the differences in bearings between sun and moon at the UTCs you have listed. Do the same thing for the sun and Venus, if you are considering these sights. These differences in bearings will be the angles between the LOPS if you took sights of them at about that same time.

For example: if the height of the sun were 29° at a bearing of 083°, and at this same time the height of the Moon was 55° at a bearing of 101°, then if you did a sight of each and plotted the LOPs, these LOPs would intersect at an angle of 101°- 83° = 18°. This intersection angle at this time is slightly too narrow for a good fix. We should look for times of day, if they exist, when this intersection angle is at least 30° or so.

The Star Finder Book

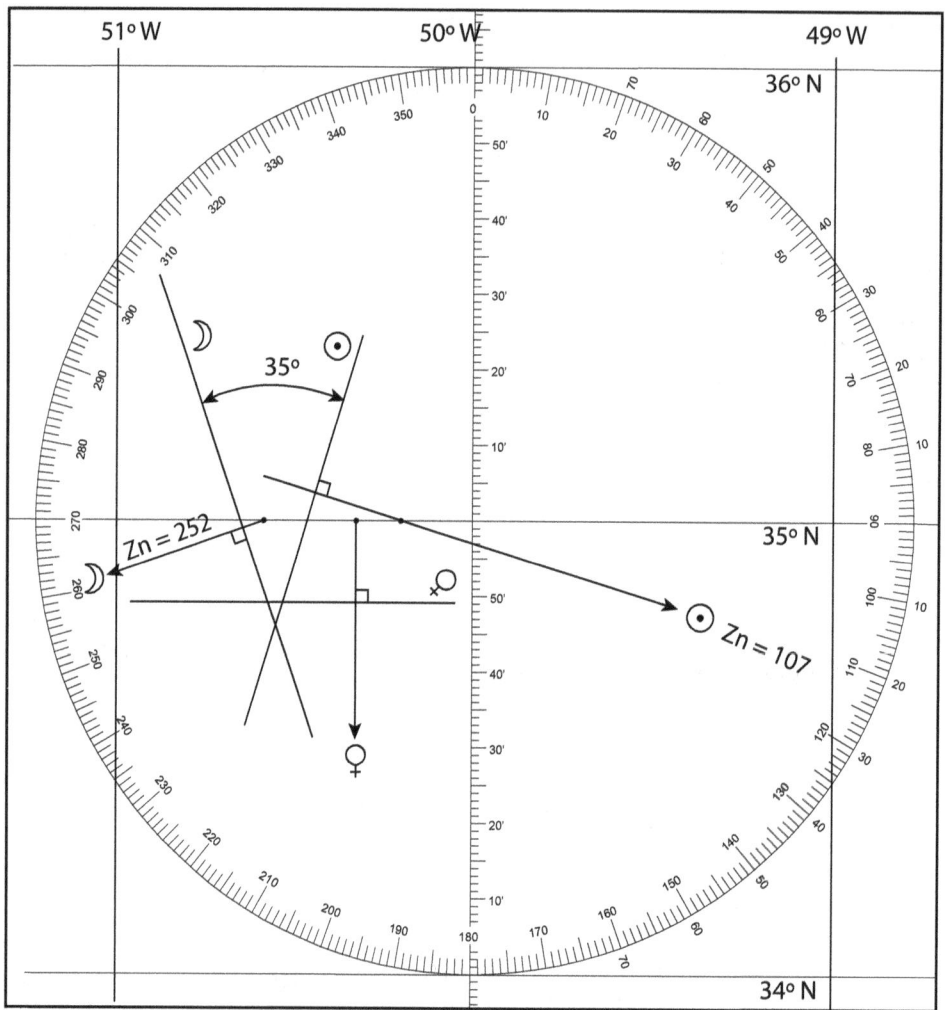

Figure 6-1 *A hypothetical sun-moon-Venus fix during daylight hours on July 14th, 1982. The optimum time to do such a fix is determined in Example 6-1. Note how relative bearings translate into LOP intersection angles—the values shown here are listed in Table 6-1.*

Figuring the LOP intersection angle from the bearings, however, is not always a simple subtraction as illustrated above. The following procedure, however, will get the right intersection angle without plotting or further reasoning.

To Find Lop Intersection Angles From Bearings

Always subtract the smaller bearing from the larger one, and then proceed according to the result:

(1) If the result is less than 90°, you are finished.

(2) If the result is greater than 90° but less than 180°, then subtract it from 180° to get the proper intersection.

(3) If the result is greater than 180°, subtract 180° from it, and go to step (1).

Follow through a few of the cases in Example 6-1, to see how this procedure works. The reasoning becomes clear if you sketch a few LOPs from Table 6-1 using arbitrary altitude intercepts ("a-values") and compare the intersection angles with the azimuths used for plotting. This is illustrated in Figure 6-1.

Example 6-1
Optimizing Sun-Moon & Sun-Venus Sights

July 14th, 1982, sailing in latitude 35° N at longitude 50° 10' W, we want to prepare for sun-moon sights and also check the possibility of sun-Venus sights. The age of the moon on this date is 23 days (see Appendix); Table 3-3 tells us there is chance for sun-moon sights near mid morning. *Nautical Almanac* times (LMT) of rising and setting for July 14th are shown below.

The sun comes up at about 0500 and sets at about 1900 LMT. The moon crosses the meridian at about 0600 and sets at about noon. A listed moonrise time of over 2400 means there is no moonrise on July 14th; the time listed (2419) is for the following day (0019 on July 15). Checking back, the listed moonrise time for the previous day, July 13th, is 2346, so we see that moonrise is just before midnight, putting it just above the horizon as July 14th begins. Venus crosses the meridian about 2 hours before the sun, so we estimate that it will rise and set about 2 hours before the sun. Since the moon sets at noon, we can see that Table 3-3 must be about right—we only have the sun and moon together during morning hours. Venus will be above the horizon throughout this period, so there may be a chance for this sight also.

The LMT halfway between sunrise (0500) and moonset (1230) is 0845. Our DR-Lon of 50° 10' W = 3h 20m 40s, which we round to 0321. The midway UTC we will use to set up the Star Finder is UTC = 0845 + 0321 = 1206, which we round to 1200 UTC.

See the Appendix for *Nautical Almanac* Data for July 14th, 1982 at 1200 UTC.

All values have been converted to decimal degrees; the times are for latitude 35° N, and they have been converted to UTC by adding 3h 21m to the LMTs listed—for example, the listed sunrise time of 0817 UTC = 0456 LMT + 0321.

Nautical Almanac gives GHA of Aries at 1200 UTC as 112° 01.5', so LHA of Aries = 112° 01.5' – 50° 10' (DR-Lon W) = 61° 51.5' = 61.85°, and that is where we mark the white disk rim scale to correspond to 1200 UTC. Each 15° from there corresponds to 1 hour away from noon: 46.85° = 1100 UTC, 76.85° = 1300 UTC, and so forth.

When these values are used to plot the body positions on the white disk, and the blue template for 35° N is used, the heights, bearings, and LOP intersection angles listed in Table 6-1 can be found from the Star Finder. Don't expect your Star Finder to give you the exact values in Table 6-1, since the table values were calculated by computer which corrected for the motion of the bodies through the stars during this day. But your Star Finder values should be close, within a few degrees. Use this comparison as a check of the Star Finder and your accuracy in plotting and reading the scales.

Table 6-1 shows that sun and moon lines intersect at the best angles (near 90') at sunrise and again just before moonset. But at these times either the sun or the moon would be too low for a good sight. In this particular circumstance, the sun-moon intersection angles are usable throughout the time that both are above the horizon. If we consider heights above 20° or so as minimum, then the usable times for sun—moon fixes would be about 0700 to 1000, LMT.

Data for Example 6-1, 7/14/1982, DR 35°00' N, 50°10' W

	Sun	Moon	Venus			
Rising	0456	2419	(0300) estimated			
Mer. Pass.	1206	0605	0959			
Setting	1915	1231	(1700) estimated			

Body	GHA	SHA	360-SHA	Dec	Rise	Mer. Pass.	Set
Sun	358.6	246.5	113.5	N 21.7	0817	1527	2236
Moon	86.0	334.0	26.0	N 5.2	(0340)	0926	1552
Venus	30.0	278.0	82.0	N 22.2	(0621)	1320	(2021)

Table 6-1. Precomputed LOP Intersection Angles Throughout the Day

		Sun (S)		Moon (M)		Venus (V)		LOP Intersection Angles		
LMT	GMT	Hc	Zn	Hc	Zn	Hc	Zn	S-M	S-V	M-V
0530	0851	5	67	58	164	31	83	83	16	81
0600	0921	11	71	59	177	37	87	74	16	90
0630	0951	17	75	58	191	43	91	64	16	80
0700	1021	23	79	56	204	49	96	55	17	72
0730	1051	29	83	54	215	55	101	47	19	66
0800	1121	35	87	50	225	61	108	42	22	63
0830	1151	41	91	45	233	67	117	38	27	64
0900	1221	47	95	40	240	72	131	35	35	70
0930	1251	54	101	35	247	76	151	34	51	85
1000	1321	60	107	30	252	77	181	35	72	71
1030	1351	65	116	24	257	76	209	39	87	47
1100	1421	70	128	18	261	72	230	47	78	32
1130	1451	75	147	12	266	66	243	61	84	23
1200	1521	77	174	7	270	61	252	84	78	18
1230	1551	76	203	1	274	55	259	70	55	15
1300	1621	72	225			49	264		39	
1330	1651	67	240			43	269		29	
1400	1721	62	250			3,6	273		23	
1430	1751	56	257			30	277		20	
1500	1821	50	263			24	281		18	
1530	1851	44	268			18	285		17	
1600	1921	38	272			12	289		17	
1630	1951	31	276			7	292		17	
1700	2021	25	280			1	297		17	
1730	2051	19	284							
1800	2121	13	287							
1830	2151	8	291							
1900	2221	2	295							

Sun-Venus intersections are good from 0900 until about 1330 LMT. At this latitude and date, the sun is not too high for a good sight near midday, so sun-Venus fixes might be possible throughout this period; the optimum times are between 1000 and 1200 LMT.

Of particular interest is, between 0900 and 1000 LMT, there is a chance for a good three-body fix, as shown in Figure 6-1. This is the only time of day, on this date and at this latitude, that this might be possible. The 2102-D Star Finder is a convenient way to discover this.

Again, as with precomputation of stars, to be certain that your predicted heights are right for doing the sights, you can work through actual sight reductions for the time you plan to do the sights. For sun-moon sights this extra precision is not needed, but it will be impossible to find Venus during daylight if it does not fall into view at your predicted height.

In practice this procedure is not as complicated as it might appear from this example. First, in actual navigation, one is almost solely interested in sun-moon sights, which are extremely valuable, and can likely be used on any ocean crossing. But you might sail happily around the world and never need or

Data for Example 6-2, 12/22/1982; DR 34°20' S, 40°18' W

Nautical Almanac Data for 1800 GMT, December 22 . 1982 :

Body	Dec	GHA – GHA Aries	= SHA
Sun	S 23° 26.4'	90° 20.5' – (0° 57.7')	= 89° 22.8' = 89.4°
Moon	S 08° 22.1'	7° 19. 5' – (0° 57.7')	= 6° 21. 8' = 6. 4°
Aries		0° 57.7'	

The values we need to plot the body positions on the white disk are :

	Dec	360° – SHA	= Rim Scale
Sun	S 23° 26.4' = S 23.4°	360° – 89.4°	= 270.6°
Moon	S 08° 22.1' = S 8.4°	360° – 6.4°	= 353.6°

have the opportunity for a daylight Venus sight. The usefulness of Venus during the day (or during twilight, for that matter) depends on the time of year; and its most useful season varies from year to year. For daylight Venus sights, you need accurate pre-computation, a good sextant telescope, and crystal clear skies. Any haze at all, or even the sun's glare in clear skies when Venus is nearer the sun, will obscure daytime sights of Venus.

And second, from a practical point of view, you can find an adequate time for sun-moon sights very quickly, without going through this procedure to find the very best time. To do this just plot the sun and moon on the white disk, with the rim scale labeled with watch time for your present longitude, and then check morning, midday, and afternoon heights and bearings. Table 3-3 lists the days you might have a chance for the sights. Example 6-2 illustrates this approach, with one more example of planning sun-moon sights.

Example 6-2
Picking the Time for Sun-Moon Sights

The date is December 22, 1982; our DR position is 34° 20' S, 40° 18' W. We want to check the possibility of a sun-moon fix during the day. The moon's age on this date is 7 days. The zone description of our navigation watch is +3 hr. Checking Table 3-3 we find that for moon age 7 days we have a chance for daytime sights, probably around mid afternoon.

Checking the Almanac daily page for December 22, 1982 we find that the moon rises at 1051 LMT and the sun sets at 1914 LMT. This, then, would be the time range we must check. Our longitude of 40°

Table 6-2. Sun-Moon Fixes Predicted by the Star Finder

		Moon		Sun		LOP Intersection Angles
GMT	WT	Hc	Zn	Hc	Zn	
1300	1000	—	—	64	068	—
1400	1100	3	097	75	040	57
1500	1200	15	088	77	340	72
1600	1300	26	080	68	300	40
1700	1400	37	070	57	282	32
1800	1500	47	057	45	272	35
1900	1600	57	035	33	264	49
2000	1700	62	008	20	256	68
2100	1800	60	337	8	248	82
2200	1900	52	312	—	—	—

The Star Finder Book

18' W = + 2h 41m, so the UTC range is 1051 + 0241 to 2914 + 0241 = 1332 to 2155 UTC, or about 1330 UTC to 2200 UTC. To set up the Star Finder we will use almanac data for the time about midway through the range, at UTC = 1800.

Use the above data and the red template to mark these positions on the south side of the white disk. Remember that south declinations are toward the center of the disk on the south side.

To set up the time scale on the white disk we need the LHA of Aries. At UTC = 1800, GHA Aries = 0° 57.7', so LHA Aries = 0° 57.7' − 40° 18' (+360°) = 320° 39.7' = 320.7°. Therefore the rim scale position 320.7 corresponds to 1800 UTC, and since the ZD of the watch is +3 hr, this corresponds to WT = 1500. We now mark the 320.7 position with WT = 1500, and label the rest of the scale from there. Every 15° from 320.7 equals 1 hour, as shown in Figure 6-2.

Now place the south side of the blue template for Latitude 35° on the white disk and read the heights and bearings listed in Table 6-2. The numbers in the table were read from a Star Finder, not calculated, so they are not precise, but they should be close to what you get.

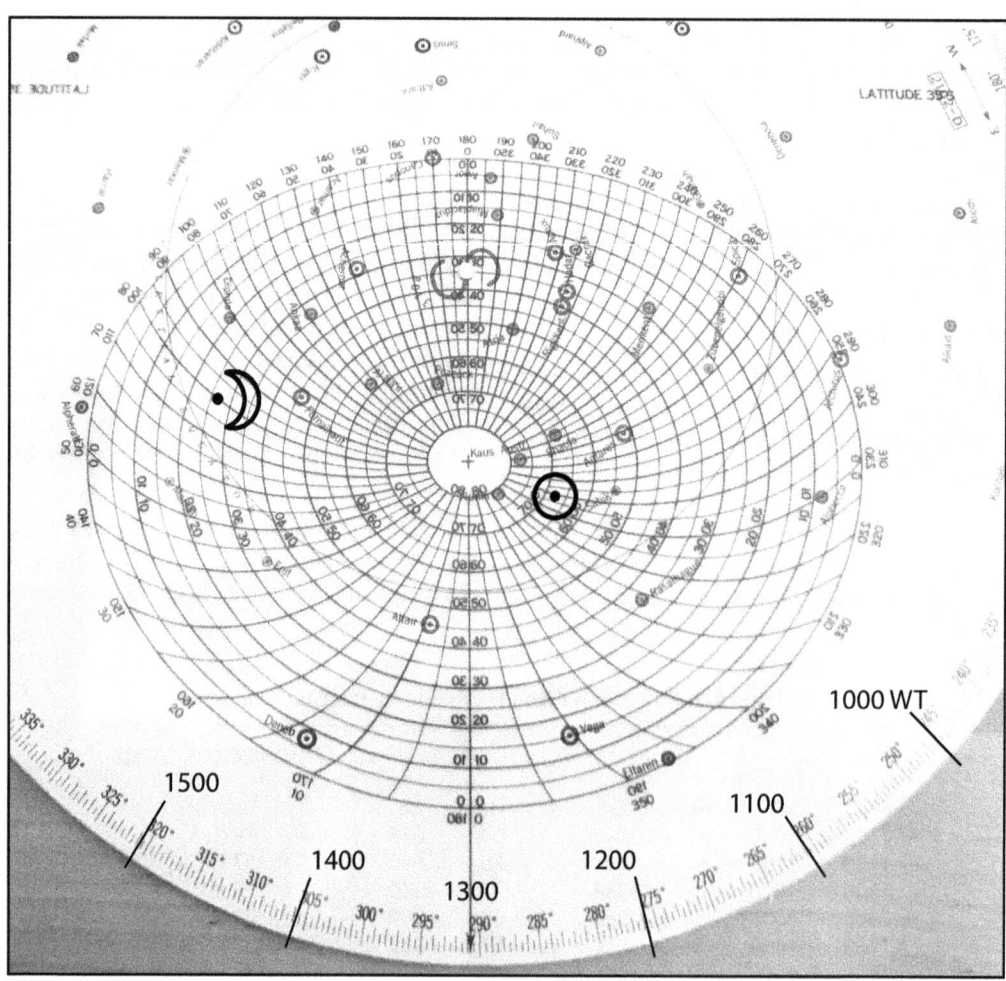

Figure 6-2 *Sun and moon plotted on the south side of the white disk with the 35° S template in place. LHA Aries = 320.7° corresponds to 1500 WT. The template here is set to 1300 WT. The moon bears 080° at a height of about 26° and the sun bears 300° at a height of about 68°. Note that LHA Aries increases to the left of the South side of the white disk.*

Note that the moon rises at about 1330 UTC (1030 WT), and the sun sets at about 2200 UTC, as they should according to the Almanac.

From Table 6-2 we can see that there is a good sun-moon fix in mid afternoon, specifically, between 1500 and 1700 WT. After 1700 the sun is too low, and before 1500 the intersection angle is near its lower limit for a strong fix, although it is probably usable throughout the afternoon. When the sun peaks (Local Apparent Noon) at about 1200 WT, the moon is too low, but 30 minutes or so after LAN a good moon sight is available. In this case you could combine it with a latitude sight at LAN for a fix, or just skip the latitude sight and do a sun-moon fix as soon as the moon is up 20° or so... assuming the moon is visible. The Star Finder will always tell us where the moon is, but it won't tell us if we can see it.

The alternative to the Star Finder method for optimizing these sights is to copy the data from Sight Reduction Tables using LHA instead of LMT. The time it takes for this method depends on the type of sight reduction tables you use. But it is not much faster, if at all. The Star Finder method offers a more direct picture of how these bodies move across the sky. It is also easier to check that you are doing it right, so there is less chance of error. This point is discussed further in Section 6.3.

6.2 A Precaution for Daylight Sights

Always remember that doing moon or Venus sights during the day is one of the rare times that celestial navigation can be potentially hazardous. Do not attempt the sights if either Venus or the moon is too close to the sun for a good fix in the first place. In other words, don't look for Venus just for the sake of seeing it in the sextant—or worse, with binoculars—when you know already that it is too close to the sun to offer a good celestial fix. If the object is too close to the sun for a fix, it is too close to the sun to look for. You will fry your retina in an instant if you ever look at the sun through the direct view of the un-shaded sextant telescope. To do this, however, you must be pointing the sextant at the sun rather than at the horizon, and this is not required for any part of these sights.

When looking for Venus or the moon with the sextant you will not have the sun shades in place, so you must be careful not to let the sun get into direct view. Set the sextant to the proper height, and only look at the horizon, not up at the sky. Keep the other eye, and the mind, on where the sun is while you look for the moon or Venus, and don't point the sextant in that direction. If the sight is usable, the moon or Venus bearing will be more than 30° away from the sun, so with simple precautions you should have no trouble. Just keep in mind the danger while doing these sights. If you stumble, or the boat rocks in some way that starts you pointing toward the sun, shut your eyes.

6.3 Star Finder versus Alternative Methods

As we have mentioned throughout, every problem the Star Finder solves can be solved by some alternative method or set of tables. But there is no one single alternative to the Star Finder for all of these problems except for a sophisticated programmable calculator—and this we can rule out as a long-term, sole solution for a small boat at sea because they don't work when they are wet and they don't like being thrown across the cabin very many times.

Restricting ourselves to convenient tables still in print, you can choose and precompute stars with Pub. 249 Vol. 1 but you must use Pub. 249 Vols. 2 and 3, or the corresponding volumes of Pub. 229, to precompute planets. Publication 249 is the most convenient for optimizing sun-moon sights because of the table organization, but for star ID after the sight Pub. 229 must be used in many cases because of the Declination limitation of Pub. 249.

On the other hand, Pub. 214—which many navigators miss—was set up nicely to do most of these jobs and it even included special star ID tables, but it is out of print. And tables like Pubs. 208 and 211, though still available, and capable of doing all these jobs, are simply too cumbersome for these specialized applications. In fact, most navigators find that these latter two sets of tables are even too cumbersome for routine navigation, despite their mathematical elegance and strong support by devotees.

But this is not the main point. The most critical job the Star Finder does is identify unknown stars after the sight. This important job can, indeed, be done by Pub. 229 for any star or planet. But I do not hesitate at all to say that the Star Finder is by far more reliable and much easier to use. The procedure for doing star ID using Pub. 229 is not simple. Just check the instructions in the tables for doing it—they start out by saying most navigators have other, preferred ways of doing this (their way of telling us to use the Star Finder). Or check anyone else's attempt (including mine, I admit) to rewrite these table instructions. Furthermore, typical table instructions do not cover the "trick cases," they usually illustrate the simple cases. Star ID with Pub. 229 is easy in some cases, and almost impossible in others.

If we are relying on celestial observations for the safe, efficient navigation of a boat, we need clear reliable methods that work in any circumstance. We do not need academic exercises or gimmicks. Since we need to learn the Star Finder for the important job of star ID, we might as well use it in other applications if we can; the procedures are all the same. This way we avoid having separate procedures or tables for these similar problems in celestial navigation.

Chapter 7.
Other Star Finder Applications

Other Applications include emergency steering and great circling sailing filler text filler. ther Applications include emergency steering and great circling sailing filler text filler.

7.1 Emergency Steering with the Star Finder

If some mishap at sea should leave you to navigate with limited instruments and tables, the Star Finder could be a valuable asset. Steering without a compass using sun and stars, for example, is easily carried out with the Star Finder alone. Pick the proper blue template to match your DR latitude and then use any known star bearing or height to mark the rim scale with time. For example, at latitude 35° South, you might spot Canopus at a height of one handwidth (about 25°) above the southeast horizon at one hour after sunset. Rotating Canopus to a height of 25° on the 35° template shows that the LHA of Aries must be about 32° at the time. Label this rim point with 1 hour after sunset and then you have a time scale to follow stars throughout the night. Note that you need only relative time from any watch, not UTC, to keep track of star bearings this way.

To improve on this, look for star pairs that have the same SHAs. These pairs will "stand up" as they cross the meridian, and you can use this observation to improve the rim scale time calibration. On very clear nights, you might also watch for bright stars rising or setting for further checks on the time scale. In the Northern Hemisphere, use Polaris to find the bearing to some bright star, and then use the bearing of that star to find LHA Aries on the Star Finder.

Then when Polaris is obscured, you have the Star Finder to go by. None of these applications require an Almanac.

If you know UTC, then you can figure GHA Aries from:

GHA Aries = 99° + (Day of the Year) × 360°/365 + (UTC) × 15°/hour.

For example, find GHA Aries at 0500 UTC on July 14th. Counting days, July 14th is day number 195 (3 x 31 + 2 x 30 + 28 + 14), so GHA Aries = 99 + 195 x 360/365 + 5.0 x 15 = 366.3° = 6.3°. Checking the Appendix we see the proper value is 6° 44' = 6.7°. The approximate formula is typically accurate to within 1°, which is good enough for emergency steering. Then from your DR-Lon figure LHA Aries and set up the Star Finder for star bearings throughout the night (each 15° on the rim scale is 1 hour).

Once you have star bearings, measure the bearing of sunrise relative to the stars (using apparent wind direction or Venus bearings for orientation during twilight). Then plot the sun on the white disk so that its bearing on the horizon is reproduced and you have sun bearings throughout the day for orientation.

The Star Finder is not accurate enough to help with finding or keeping track of position. Emergency methods of no-instrument navigation are covered in detail in the author's book *Emergency Navigation* (International Marine Publishing Company, Camden ME, 2008).

Have fun with your Star Finder. The more you use it, the more valuable it will become to your navi-

The Star Finder Book

gation and stargazing. If you have questions, comments, or suggestions about the Star Finder or this book, send them to helpdesk@starpath.com.

7.2 Great Circle Sailing

When we place a blue template on the white disk, we are locating the pole of the sky on the pivot point of the blue template. The altitude (Hc) of this point will be our latitude. The point in the center of the blue template is the geographical position (GP) of the observer, and the radial lines emanating from it are azimuth lines marking the true bearings (Zn) to the stars plotted. The angular distance on the surface of the earth between the GP of a star and the GP of the observer is 90° minus Hc of the star above the horizon. Sixty times this angle is the great circle (GC) distance in nmi between the two points.

Let's consider a hypothetical case of being located at a waypoint we call Deneb, and wanting the great circle distance and initial heading to a waypoint called Hamal, as shown in Figure 7-1. These waypoints happen to be the fixed positions of two navigational stars that are plotted on the white disk shown in Figure 7-2.

We chose Deneb for this example because its declination (N 45° 21.2') nearly matches the Lat 45 N blue template of the Star Finder. We chose Hamal more or less randomly.

Thus we imagine the earth not rotating and we are at the GP of Deneb, meaning it is directly overhead, 90° above the horizon, and we want the initial GC heading and distance to Hamal. With the disk rotated so that Deneb is in the center of the blue diagram, the blue lines emanating from that point are all great circles, so we just find the one that goes from Deneb to Hamal, and read that true bearing on the rim of the blue template, which corresponds to the horizon as viewed from the GP of Deneb.

In this case, we see Zn to Hamal is about 078 or 079 T, and Hc of Hamal is about halfway between 20° and 25° above the horizon. Thus the GC distance between them is 90° − 22.5° = 67.5°, and each degree is 60 nmi, so the GC distance we read is 4050 nmi.

Thus, in this example, we get from the Star Finder an initial heading of 079 ±0.5 compared to correct value of 078.5 and a star finder distance of 4050 compared to a correct value of 4063.5.

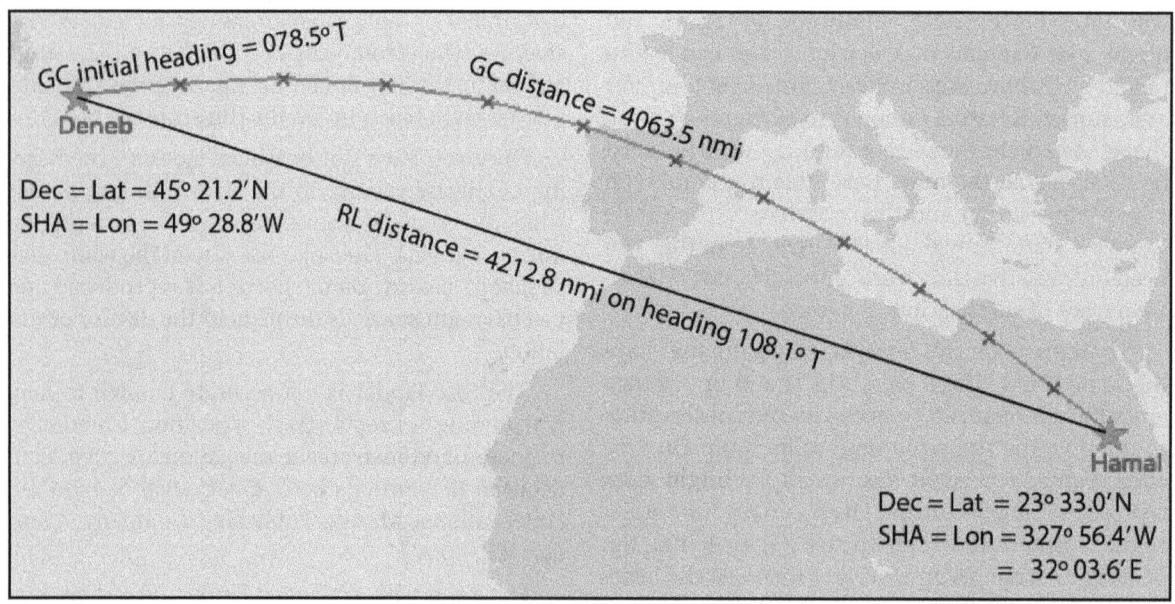

Figure 7-1 *Hypothetical Deneb-Hamal great circle route compared with rhumb line. Accurate great circle and rhumb line solutions can be computed at www.starpath.com/calc.*

The GC heading is a whopping 30° north of the RL route in this example, so this could have a major impact on navigation decisions. We don't care so much that one route is shorter than the other, even when this difference is large as in this case, because we are dominated by wind and rarely can make good such routes. But knowing that 078 is just as good or even better than 108 gives us some freedom in planning what to do in local winds.

In the real world, our initial latitude will not coincide with a template value, so we have to improvise the process.

The Procedure

There is a video link at the end showing the plotting.

(1) On the N side of the white disk, draw a thin line to mark the departure meridian going from 0° on the rim to center of the centerpin.

(2) Use the red template scales to plot your departure Lat, along the departure meridian, on the white disk. The celestial equator on the white disk is equivalent to the equator for this plotting. It is likely best to use dividers on the red template to get the right lat spacing, and then transfer that to the white disk. All

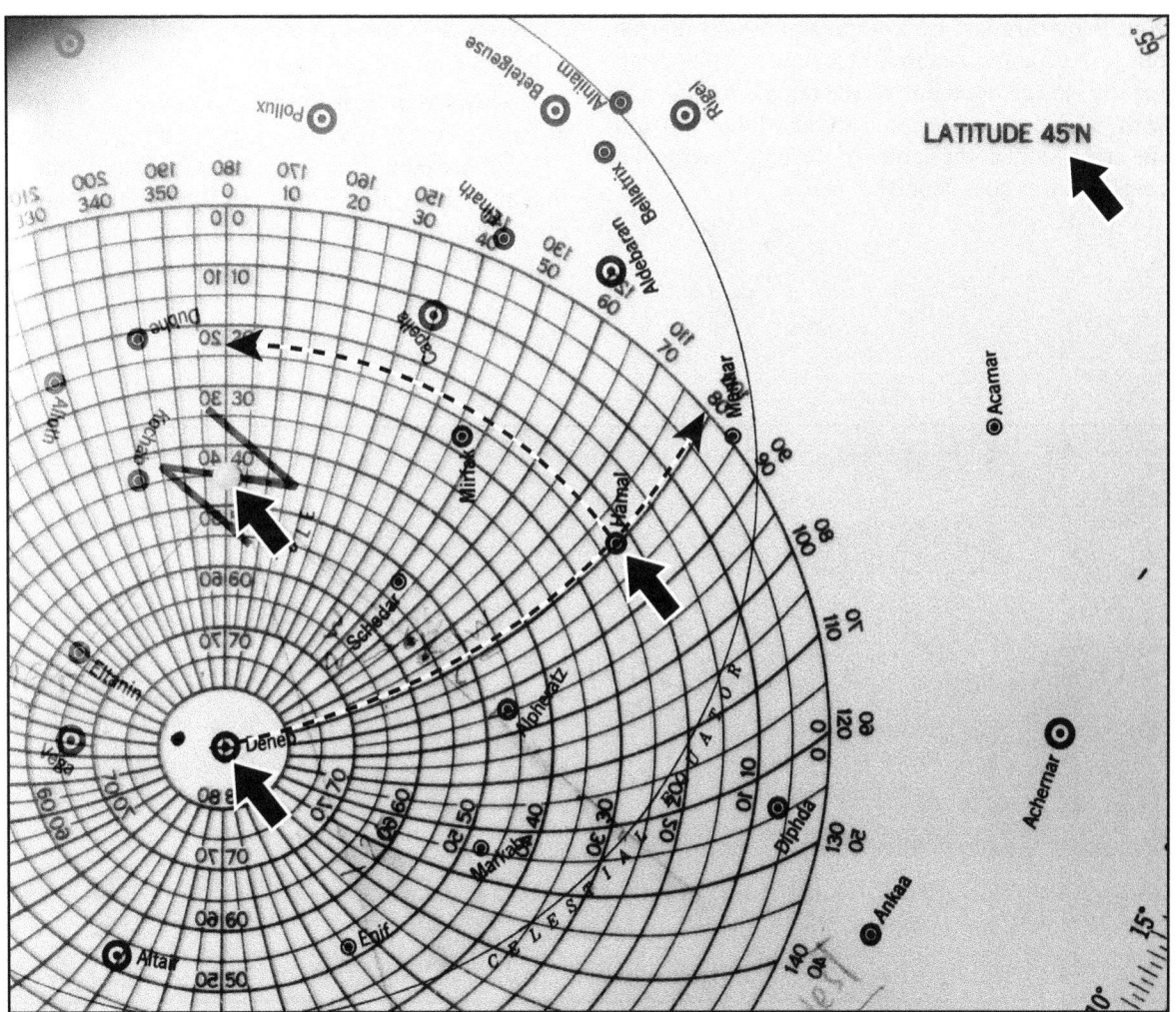

Figure 7-2 *Hypothetical Deneb-Hamal great circle initial heading and distance as found with the Star Finder.*

of this plotting should be done as carefully as possible as the scales involved are all compressed.

(3) Figure the Lon difference (dLon) between departure and arrival, and note if arrival is west or east of departure. If arrival is to the east, the arrival meridian is just dLon to the right of 0° on the rim. If arrival is to the west, the arrival meridian is located at 360 − dLon to the left of 0°. Again, draw in the arrival meridian carefully as noted previously.

(4) Set dividers to arrival Lat on the red disk and then plot it on the arrival meridian.

(5) Find the blue template with the closest Lat to your departure Lat. Do not put it on the centerpin, but instead move it above or below the pin (keeping the blue arrowed line on the template coinciding with the departure meridian on the white disk) until the cross hairs at the center of the blue diagram are precisely over your departure point.

(6) Then hold the template in place and carefully read the Hc and Zn to your arrival point, interpolating as best you can. The initial GC heading is Zn; the total GC distance is approximately (90 − Hc) × 60 nmi.

Here is a video link showing two examples:

https://youtu.be/7kQCmxqd94s

Example 1. West Coast of US at 45° N, 125° W to Japan 38° N, 142° E. For this route the GC distance is 3962.1 nmi, with an initial heading of 300.6° T. This heading is 36.3° north of the RL heading of 264.3. The GC distance is 247.1 nmi shorter than the RL distance of 4209.2 nmi. Star Finder solution is 302 T and 3900 nmi.

Example 2. Exit of the Strait of Juan de Fuca at 48° N, 125° W to HI at 21.5° N, 157° W. This is a GC distance of 2210.3, which is just 14.7 nmi shorter than the RL distance of 2225.0. The initial GC heading of 235.3° T is 10.9° north of the RL heading of 224.4. Star finder solution is 235 T and 2100 nmi.

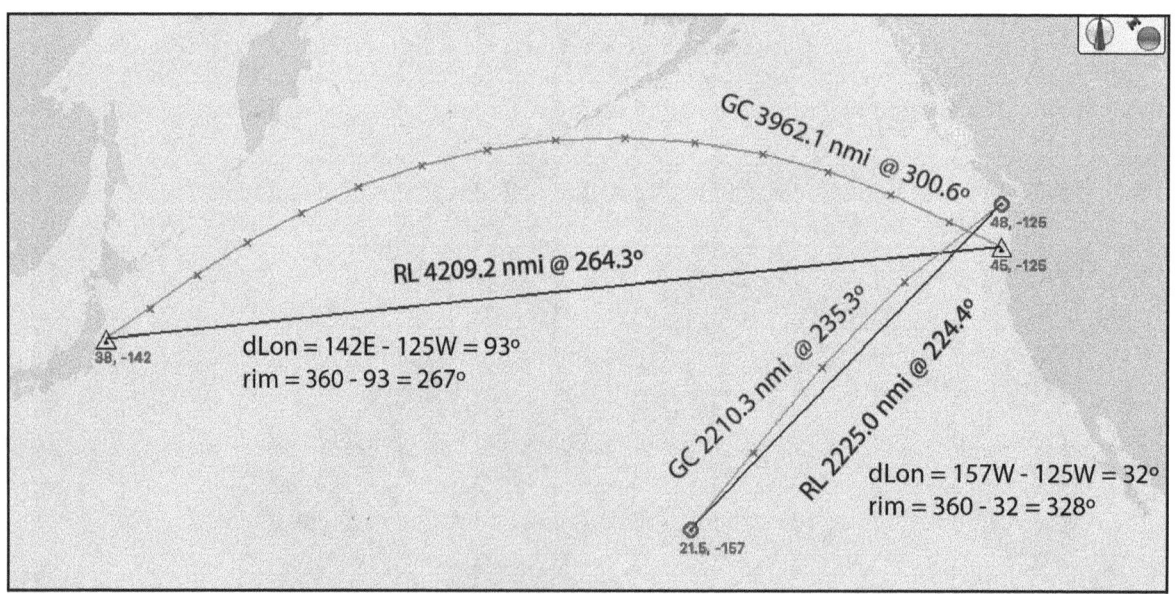

Figure 7-3 *Two sample great circle solutions worked with the Star Finder. Screen capture from OpenCPN. Numerical values shown were computed at www.starpath.com/calc.*

Bibliography

The American Practical Navigator, originally by Nathaniel Bowditch (NGA Pub. No.9, Washington, D.C., updated periodically). The encyclopedia of marine navigation. Celestial navigation topics have consistently diminished in content since the 1977 edition. The latest and most older editions are online in full.

Star Names Their Lore and Meaning, Richard H. Allen (Dover Publications, Inc., New York, 1963). If a star has a name this book will tell you what it is and where it came from. Many stars have names that do not appear on popular star charts or lists. A scholarly treatment, but still interesting reading, now available in full online.

Practical Astronomy With Your Calculator, Third Edition, Peter Duffett-Smith (Cambridge University Press, New York, 1988). Lists and explains the equations used in calculating the almanac data If you have an interest in positional astronomy or care to program a calculator or personal computer for almanac predictions this is an extremely valuable reference.

North Star to Southern Cross, Will Kyselka and Ray Lanterman (The University Press of Hawaii, Honolulu, 1976). A short book on the stars with convenient star maps for each month. Covers the constellations and related mythologies with a nice summary of astronomy. An enjoyable companion for an ocean voyage.

Observer's Handbook, (Royal Astronomical Society of Canada, www.rasc.ca/handbook). Annual handbook on moon and planet locations and much other data for amateur astronomers. Lists star data for the 286 brightest stars in the sky. Excellent treatment of eclipses and other astronomical events like meteor showers and so forth, presented in a convenient calendar format. More for astronomy interests than for navigation.

"The Heavens," a large durable paper star map produced by the National Geographic Society. See their speciality maps section.

The Nautical Almanac (United States Naval Observatory, Washington, D.C., annual).

Commercial Edition of the Nautical Almanac (Paradise Cay Publications, Arcada, CA, annual.) A reproduction of the USNO *Nautical Almanac*.

Long Term Almanac for Sun and Stars, Geoffrey Kolbe (Starpath Publications, Seattle, 2008). Provides GHA and declination until 2050, along with altitude correction tables and a complete copy of the NAO Sight Reduction Tables. A stand alone celestial navigation solution.

Emergency Navigation, Second Edition, David Burch, (International Marine Publishing Company, Camden ME, 2008). Covers apparent star motions and steering by the stars without instruments among other topics of no-instrument navigation.

Celestial Navigation: A Complete Home Study Course, Second Edition, David Burch (Starpath Publications, Seattle, 2019).

Hawaii by Sextant: An In-depth Exercise in Celestial Navigation Using Real Sextant Sights and Logbook Entries, David Burch and Stephen Miller (Starpath Publications, Seattle, 2017). An excellent source of data to practice star finder usage.

Appendix

Almanac Data

Nautical Almanac daily pages for July 12, 13, and 14 of 1982. Most text examples use these data, though the times are not at the precise hours tabulated. To verify the data for specific examples, take the hours part from these tables and then make the minutes and seconds corrections using the Increments and Corrections Tables from any *Nautical Almanac*—these corrections do not change from year to year.

In the text Example 4-1, for example, to check the GHA Aries on July 14th 1982 at 05h 12m 20s, get the 5h part (6° 44.3') from these tables and the 12m 20s part (3° 5.5') from an Almanac of any year and sum them to verify the final value of 9° 49.8'.

Following these pages are the Star Maps from the *Nautical Almanac* described in Chapters 2 and 3.

Directions From the Stars

Below: POINTING MINTAKA (LEADING STAR OF ORION'S BELT) BACKWARD TO THE HORIZON TO FIND EAST. The rising angle to use is 90° minus latitude. The figure shows this being done from latitude 35° N (rising angle 55°) at 2.5 hours after Mintaka rose. The method works from most latitudes for 2 to 3 hours past the rising time. The figure is roughly to scale, showing that the same method applied 3.5 hours after rising would give an error of about 10°. Hourly positions of Mintaka, after rising, are marked with circled Xs.

Above: ORION ON THE HORIZON. From any point on earth, at any time of night, Orion's belt always rises due east and sets due west. The Seven Sisters, Pleiades, lead the chase of nearby stars. Taurus the Bull follows, fighting off Orion whose faithful hunting dogs, Sirius and Procyon, trail close behind. Betelgeuse, at the base of Orion's raised arm, and Aldebaran, at the eye of the Bull, are brilliant red giant stars.

Adapted from Emergency Navigation, by David Burch, (McGraw-Hill, 1986, 2008). See related discussion in Chapter 7 of this book

1982 JULY 12, 13, 14 (MON., TUES., WED.)

Appendix

G.M.T.	SUN G.H.A.	SUN Dec.	MOON G.H.A.	MOON v	MOON Dec.	MOON d	MOON H.P.	Lat.	Twilight Naut.	Twilight Civil	Sunrise	Moonrise 12	Moonrise 13	Moonrise 14	Moonrise 15
d h	° '	° '	° '	'	° '	'	'	°	h m	h m	h m	h m	h m	h m	h m
12 00	178 38.1	N22 02.8	293 52.4	13.7	S 7 16.8	11.7	56.1	N 72	▢	▢	▢	23 40	23 22	23 03	22 39
01	193 38.0	02.5	308 25.1	13.7	7 05.1	11.8	56.2	N 70	▢	▢	▢	23 37	23 25	23 13	23 00
02	208 37.9	02.2	322 57.8	13.7	6 53.3	11.8	56.2	68	▢	▢	01 24	23 34	23 28	23 22	23 16
03	223 37.8	·· 01.8	337 30.5	13.6	6 41.5	11.9	56.2	66	////	////	02 08	23 31	23 30	23 29	23 29
04	238 37.8	01.5	352 03.1	13.7	6 29.6	11.9	56.2	64	////	////	02 37	23 29	23 32	23 36	23 41
05	253 37.7	01.1	6 35.8	13.7	6 17.7	11.9	56.3	62	////	00 25	02 59	23 27	23 34	23 41	23 50
06	268 37.6	N22 00.8	21 08.5	13.7	S 6 05.8	12.0	56.3	60	////	01 37	03 17	23 26	23 35	23 45	23 59
07	283 37.5	00.4	35 41.2	13.7	5 53.8	12.0	56.3	N 58	////	02 10	03 31	23 24	23 37	23 50	24 06
08	298 37.4	22 00.1	50 13.9	13.6	5 41.8	12.1	56.4	56	00 24	02 34	03 44	23 23	23 38	23 54	24 12
M 09	313 37.4	21 59.8	64 46.5	13.7	5 29.7	12.1	56.4	54	01 29	02 53	03 55	23 22	23 39	23 57	24 18
O 10	328 37.3	59.4	79 19.2	13.6	5 17.6	12.1	56.4	52	02 00	03 09	04 05	23 21	23 40	24 00	00 00
N 11	343 37.2	59.1	93 51.8	13.7	5 05.5	12.2	56.4	50	02 22	03 22	04 25	23 20	23 41	24 03	00 03
D 12	358 37.1	N21 58.7	108 24.5	13.6	S 4 53.3	12.1	56.5	45	03 02	03 49	04 42	23 18	23 43	24 09	00 09
A 13	13 37.0	58.4	122 57.1	13.6	4 41.2	12.3	56.5	N 40	03 30	04 10	04 56	23 16	23 44	24 14	00 14
Y 14	28 37.0	58.0	137 29.7	13.6	4 28.9	12.2	56.5	35	03 51	04 27	05 08	23 15	23 46	24 19	00 19
15	43 36.9	·· 57.7	152 02.3	13.6	4 16.7	12.3	56.5	30	04 08	04 41	05 28	23 13	23 47	24 23	00 23
16	58 36.8	57.3	166 34.9	13.6	4 04.4	12.3	56.6	20	04 35	05 04	05 46	23 11	23 49	24 29	00 29
17	73 36.7	57.0	181 07.5	13.6	3 52.1	12.4	56.6	N 10	04 56	05 23	06 02	23 09	23 51	24 36	00 36
								0	05 14	05 40	06 02	23 07	23 53	24 41	00 41
18	88 36.6	N21 56.6	195 40.1	13.5	S 3 39.7	12.4	56.6	S 10	05 30	05 56	06 18	23 05	23 55	24 47	00 47
19	103 36.6	56.3	210 12.6	13.5	3 27.3	12.4	56.7	20	05 44	06 12	06 35	23 03	23 57	24 53	00 53
20	118 36.5	55.9	224 45.1	13.5	3 14.9	12.4	56.7	30	05 59	06 29	06 55	23 01	24 00	00 00	01 01
21	133 36.4	·· 55.6	239 17.6	13.5	3 02.5	12.5	56.7	35	06 07	06 39	07 06	23 00	24 01	00 01	01 05
22	148 36.3	55.2	253 50.1	13.5	2 50.0	12.4	56.8	40	06 16	06 49	07 19	22 58	24 03	00 03	01 10
23	163 36.3	54.9	268 22.6	13.5	2 37.6	12.5	56.8	45	06 25	07 02	07 35	22 57	24 05	00 05	01 15
13 00	178 36.2	N21 54.5	282 55.1	13.4	S 2 25.1	12.6	56.8	S 50	06 35	07 16	07 54	22 54	24 07	00 07	01 22
01	193 36.1	54.2	297 27.5	13.4	2 12.5	12.5	56.8	52	06 40	07 23	08 02	22 54	24 08	00 08	01 25
02	208 36.0	53.8	311 59.9	13.4	2 00.0	12.6	56.9	54	06 45	07 30	08 12	22 52	24 09	00 09	01 28
03	223 36.0	·· 53.4	326 32.3	13.3	1 47.4	12.6	56.9	56	06 50	07 38	08 24	22 51	24 10	00 10	01 32
04	238 35.9	53.1	341 04.6	13.4	1 34.8	12.6	56.9	58	06 56	07 47	08 37	22 50	24 12	00 12	01 36
05	253 35.8	52.7	355 37.0	13.3	1 22.2	12.7	57.0	S 60	07 03	07 58	08 52	22 49	24 13	00 13	01 41

G.M.T.	SUN G.H.A.	SUN Dec.	MOON G.H.A.	MOON v	MOON Dec.	MOON d	MOON H.P.	Lat.	Sunset	Twilight Civil	Twilight Naut.	Moonset 12	Moonset 13	Moonset 14	Moonset 15
06	268 35.7	N21 52.4	10 09.3	13.3	S 1 09.5	12.6	57.0	°	h m	h m	h m	h m	h m	h m	h m
07	283 35.6	52.0	24 41.6	13.2	0 56.9	12.7	57.0								
08	298 35.6	51.7	39 13.8	13.2	0 44.2	12.7	57.1								
T 09	313 35.5	·· 51.3	53 46.0	13.2	0 31.5	12.7	57.1								
U 10	328 35.4	50.9	68 18.2	13.2	0 18.8	12.7	57.1								
E 11	343 35.3	50.6	82 50.4	13.1	S 0 06.1	12.8	57.2	N 72	▢	▢	▢	09 33	11 28	13 26	15 35
S 12	358 35.3	N21 50.2	97 22.5	13.1	N 0 06.7	12.7	57.2	N 70	▢	▢	▢	09 41	11 28	13 18	15 16
D 13	13 35.2	49.9	111 54.6	13.1	0 19.4	12.7	57.2	68	▢	▢	▢	09 48	11 28	13 12	15 01
A 14	28 35.1	49.5	126 26.7	13.0	0 32.2	12.8	57.3	66	22 43	////	////	09 53	11 28	13 06	14 50
Y 15	43 35.0	·· 49.1	140 58.7	13.0	0 45.0	12.8	57.3	64	22 01	////	////	09 58	11 28	13 02	14 40
16	58 35.0	48.8	155 30.7	12.9	0 57.8	12.8	57.3	62	21 32	23 34	////	10 02	11 28	12 58	14 31
17	73 34.9	48.4	170 02.6	12.9	1 10.6	12.8	57.4	60	21 11	22 32	////	10 05	11 29	12 54	14 24
18	88 34.8	N21 48.0	184 34.5	12.9	N 1 23.4	12.8	57.4	N 58	20 54	21 59	////	10 08	11 29	12 51	14 18
19	103 34.8	47.7	199 06.4	12.8	1 36.2	12.8	57.4	56	20 39	21 36	23 37	10 11	11 29	12 49	14 12
20	118 34.7	47.3	213 38.2	12.8	1 49.0	12.8	57.4	54	20 27	21 17	22 40	10 13	11 29	12 46	14 07
21	133 34.6	·· 46.9	228 10.0	12.8	2 01.8	12.9	57.5	52	20 16	21 02	22 10	10 15	11 29	12 44	14 03
22	148 34.5	46.6	242 41.8	12.7	2 14.7	12.8	57.5	50	20 06	20 48	21 48	10 17	11 29	12 42	13 59
23	163 34.5	46.2	257 13.5	12.6	2 27.5	12.8	57.5	45	19 46	20 22	21 09	10 22	11 29	12 38	13 50
14 00	178 34.4	N21 45.8	271 45.1	12.6	N 2 40.3	12.9	57.6	N 40	19 29	20 01	20 41	10 25	11 29	12 34	13 42
01	193 34.3	45.4	286 16.7	12.6	2 53.2	12.8	57.6	35	19 15	19 44	20 20	10 28	11 29	12 31	13 36
02	208 34.2	45.1	300 48.3	12.5	3 06.0	12.9	57.6	30	19 03	19 30	20 03	10 31	11 29	12 29	13 31
03	223 34.2	·· 44.7	315 19.8	12.4	3 18.9	12.8	57.7	20	18 43	19 07	19 34	10 36	11 29	12 24	13 21
04	238 34.1	44.3	329 51.2	12.5	3 31.7	12.8	57.7	N 10	18 26	18 48	19 15	10 40	11 29	12 20	13 13
05	253 34.0	44.0	344 22.7	12.3	3 44.5	12.9	57.7	0	18 09	18 32	18 57	10 44	11 29	12 16	13 05
06	268 34.0	N21 43.6	358 54.0	12.3	N 3 57.4	12.8	57.8	S 10	17 53	18 16	18 42	10 47	11 29	12 12	12 58
07	283 33.9	43.2	13 25.3	12.3	4 10.2	12.8	57.8	20	17 36	18 00	18 27	10 51	11 29	12 08	12 49
W 08	298 33.8	42.8	27 56.6	12.2	4 23.0	12.9	57.9	30	17 17	17 42	18 12	10 56	11 29	12 03	12 40
E 09	313 33.8	·· 42.5	42 27.8	12.1	4 35.9	12.8	57.9	35	17 05	17 33	18 04	10 58	11 29	12 01	12 35
D 10	328 33.7	42.1	56 58.9	12.1	4 48.7	12.8	57.9	40	16 52	17 22	17 56	11 01	11 29	11 58	12 29
N 11	343 33.6	41.7	71 30.0	12.0	5 01.5	12.8	58.0	45	16 37	17 10	17 47	11 04	11 29	11 54	12 21
E 12	358 33.5	N21 41.3	86 01.0	12.0	N 5 14.3	12.7	58.0	S 50	16 18	16 55	17 36	11 08	11 29	11 50	12 13
S 13	13 35.5	41.0	100 32.0	11.9	5 27.0	12.8	58.0	52	16 09	16 49	17 32	11 09	11 29	11 48	12 09
D 14	28 33.4	40.6	115 02.9	11.9	5 39.8	12.8	58.1	54	15 59	16 41	17 27	11 12	11 29	11 46	12 05
A 15	43 33.3	·· 40.2	129 33.8	11.8	5 52.6	12.7	58.1	56	15 48	16 33	17 21	11 14	11 29	11 44	12 00
Y 16	58 33.3	39.8	144 04.6	11.7	6 05.3	12.7	58.1	58	15 35	16 24	17 15	11 17	11 29	11 41	11 55
17	73 33.2	39.4	158 35.3	11.6	6 18.0	12.7	58.2	S 60	15 20	16 14	17 09	11 19	11 29	11 38	11 49

18	88 33.1	N21 39.1	173 05.9	11.6	N 6 30.7	12.7	58.2	Day	SUN Eqn. of Time 00ʰ	SUN Eqn. of Time 12ʰ	SUN Mer. Pass.	MOON Mer. Pass. Upper	MOON Mer. Pass. Lower	Age	Phase
19	103 33.1	38.7	187 36.5	11.6	6 43.4	12.7	58.2	d	m s	m s	h m	h m	h m		
20	118 33.0	38.3	202 07.1	11.4	6 56.1	12.7	58.3	12	05 28	05 31	12 06	04 33	16 55	21	◐
21	133 32.9	·· 37.9	216 37.5	11.5	7 08.8	12.6	58.3	13	05 35	05 39	12 06	05 18	17 41	22	
22	148 32.9	37.5	231 07.9	11.4	7 21.4	12.6	58.4	14	05 42	05 46	12 06	06 05	18 29	23	
23	163 32.8	37.1	245 38.3	11.2	7 34.0	12.6	58.4								
	S.D. 15.8	d 0.4	S.D. 15.4		15.6		15.8								

63

The Star Finder Book 1982 JULY 12, 13, 14 (MON., TUES., WED.)

G.M.T.	ARIES G.H.A.	VENUS −3.3 G.H.A.	Dec.	MARS +0.6 G.H.A.	Dec.	JUPITER −1.7 G.H.A.	Dec.	SATURN +1.0 G.H.A.	Dec.	STARS Name	S.H.A.	Dec.
12 00	289 33.7	210 46.8 N21	53.0	93 04.8 S 7	29.4	80 31.6 S10	37.3	93 54.3 S 3	59.1	Acamar	315 36.4	S40 22.3
01	304 36.2	225 46.1	53.3	108 06.1	29.9	95 33.9	37.4	108 56.7	59.1	Achernar	335 44.3	S57 19.3
02	319 38.6	240 45.4	53.7	123 07.4	30.4	110 36.3	37.4	123 59.1	59.2	Acrux	173 36.0	S63 00.3
03	334 41.1	255 44.6 ··	54.0	138 08.8 ··	30.9	125 38.7 ··	37.5	139 01.5 ··	59.2	Adhara	255 31.5	S28 56.8
04	349 43.6	270 43.9	54.3	153 10.1	31.4	140 41.0	37.5	154 03.8	59.3	Aldebaran	291 16.8	N16 28.4
05	4 46.0	285 43.2	54.7	168 11.5	31.9	155 43.4	37.6	169 06.2	59.3			
06	19 48.5	300 42.4 N21	55.0	183 12.8 S 7	32.4	170 45.8 S10	37.6	184 08.6 S 3	59.4	Alioth	166 41.4	N56 03.7
07	34 51.0	315 41.7	55.3	198 14.2	32.9	185 48.1	37.7	199 11.0	59.4	Alkaid	153 17.4	N49 24.4
08	49 53.4	330 40.9	55.6	213 15.5	33.4	200 50.5	37.7	214 13.4	59.5	Al Na'ir	28 12.9	S47 02.6
M 09	64 55.9	345 40.2 ··	56.0	228 16.9 ··	34.0	215 52.9 ··	37.8	229 15.7 ··	59.5	Alnilam	276 10.7	S 1 12.7
O 10	79 58.3	0 39.5	56.3	243 18.2	34.5	230 55.2	37.8	244 18.1	59.6	Alphard	218 19.6	S 8 34.9
N 11	95 00.8	15 38.7	56.6	258 19.5	35.0	245 57.6	37.9	259 20.5	59.6			
D 12	110 03.3	30 38.0 N21	56.9	273 20.9 S 7	35.5	261 00.0 S10	37.9	274 22.9 S 3	59.7	Alphecca	126 30.8	N26 46.6
A 13	125 05.7	45 37.2	57.3	288 22.2	36.0	276 02.3	38.0	289 25.2	59.7	Alpheratz	358 07.9	N28 59.4
Y 14	140 08.2	60 36.5	57.6	303 23.6	36.5	291 04.7	38.0	304 27.6	59.7	Altair	62 30.9	N 8 49.3
15	155 10.7	75 35.8 ··	57.9	318 24.9 ··	37.0	306 07.0 ··	38.1	319 30.0 ··	59.8	Ankaa	353 38.8	S42 23.9
16	170 13.1	90 35.0	58.2	333 26.2	37.5	321 09.4	38.1	334 32.4	59.8	Antares	112 55.0	S26 23.6
17	185 15.6	105 34.3	58.5	348 27.6	38.0	336 11.8	38.2	349 34.8	59.9			
18	200 18.1	120 33.5 N21	58.9	3 28.9 S 7	38.5	351 14.1 S10	38.2	4 37.1 S 3	59.9	Arcturus	146 17.2	N19 16.6
19	215 20.5	135 32.8	59.2	18 30.2	39.1	6 16.5	38.2	19 39.5	4 00.0	Atria	108 17.7	S68 59.9
20	230 23.0	150 32.1	59.5	33 31.6	39.6	21 18.9	38.3	34 41.9	00.0	Avior	234 28.4	S59 27.2
21	245 25.4	165 31.3	21 59.8	48 32.9 ··	40.1	36 21.2 ··	38.3	49 44.3 ··	00.1	Bellatrix	278 57.7	N 6 20.1
22	260 27.9	180 30.6	22 00.1	63 34.3	40.6	51 23.6	38.4	64 46.6	00.1	Betelgeuse	271 27.2	N 7 24.3
23	275 30.4	195 29.8	00.4	78 35.6	41.1	66 25.9	38.4	79 49.0	00.2			
13 00	290 32.8	210 29.1 N22	00.7	93 36.9 S 7	41.6	81 28.3 S10	38.5	94 51.4 S 4	00.2	Canopus	264 07.2	S52 41.1
01	305 35.3	225 28.3	01.0	108 38.3	42.1	96 30.7	38.5	109 53.8	00.3	Capella	281 09.8	N45 58.7
02	320 37.8	240 27.6	01.4	123 39.6	42.6	111 33.0	38.6	124 56.1	00.3	Deneb	49 47.2	N45 13.0
03	335 40.2	255 26.8 ··	01.7	138 40.9 ··	43.1	126 35.4 ··	38.6	139 58.5 ··	00.4	Denebola	182 57.9	N14 40.4
04	350 42.7	270 26.1	02.0	153 42.3	43.7	141 37.7	38.7	155 00.9	00.4	Diphda	349 19.5	S18 04.9
05	5 45.2	285 25.4	02.3	168 43.6	44.2	156 40.1	38.7	170 03.3	00.5			
06	20 47.6	300 24.6 N22	02.6	183 44.9 S 7	44.7	171 42.5 S10	38.8	185 05.6 S 4	00.5	Dubhe	194 20.9	N61 51.1
07	35 50.1	315 23.9	02.9	198 46.3	45.2	186 44.8	38.8	200 08.0	00.6	Elnath	278 42.9	N28 35.5
08	50 52.6	330 23.1	03.2	213 47.6	45.7	201 47.2	38.9	215 10.4	00.6	Eltanin	90 56.6	N51 29.6
T 09	65 55.0	345 22.4 ··	03.5	228 48.9 ··	46.2	216 49.5 ··	38.9	230 12.8 ··	00.7	Enif	34 10.1	N 9 47.6
U 10	80 57.5	0 21.6	03.8	243 50.3	46.7	231 51.9	39.0	245 15.1	00.7	Fomalhaut	15 49.8	S29 42.4
E 11	95 59.9	15 20.9	04.1	258 51.6	47.2	246 54.3	39.0	260 17.5	00.8			
S 12	111 02.4	30 20.1 N22	04.4	273 52.9 S 7	47.7	261 56.6 S10	39.1	275 19.9 S 4	00.8	Gacrux	172 27.5	S57 01.1
D 13	126 04.9	45 19.4	04.7	288 54.2	48.3	276 59.0	39.1	290 22.3	00.9	Gienah	176 16.8	S17 26.6
A 14	141 07.3	60 18.6	05.0	303 55.6	48.8	292 01.3	39.2	305 24.6	00.9	Hadar	149 21.5	S60 17.5
Y 15	156 09.8	75 17.9 ··	05.3	318 56.9 ··	49.3	307 03.7 ··	39.2	320 27.0 ··	00.9	Hamal	328 27.6	N23 22.6
16	171 12.3	90 17.1	05.6	333 58.2	49.8	322 06.0	39.3	335 29.4	01.0	Kaus Aust.	84 14.7	S34 23.6
17	186 14.7	105 16.4	05.9	348 59.6	50.3	337 08.4	39.4	350 31.7	01.0			
18	201 17.2	120 15.6 N22	06.2	4 00.9 S 7	50.8	352 10.8 S10	39.4	5 34.1 S 4	01.1	Kochab	137 18.5	N74 14.0
19	216 19.7	135 14.9	06.4	19 02.2	51.3	7 13.1	39.5	20 36.5	01.1	Markab	14 01.7	N15 06.5
20	231 22.1	150 14.1	06.7	34 03.5	51.8	22 15.5	39.5	35 38.9	01.2	Menkar	314 39.9	N 4 01.2
21	246 24.6	165 13.4 ··	07.0	49 04.9 ··	52.4	37 17.8 ··	39.6	50 41.2 ··	01.2	Menkent	148 35.4	S36 17.1
22	261 27.1	180 12.6	07.3	64 06.2	52.9	52 20.2	39.6	65 43.6	01.3	Miaplacidus	221 45.6	S69 38.8
23	276 29.5	195 11.9	07.6	79 07.5	53.4	67 22.5	39.7	80 46.0	01.3			
14 00	291 32.0	210 11.1 N22	07.9	94 08.8 S 7	53.9	82 24.9 S10	39.7	95 48.4 S 4	01.4	Mirfak	309 14.6	N49 47.7
01	306 34.4	225 10.4	08.2	109 10.2	54.4	97 27.2	39.8	110 50.7	01.4	Nunki	76 27.2	S26 19.1
02	321 36.9	240 09.6	08.5	124 11.5	54.9	112 29.6	39.8	125 53.1	01.5	Peacock	53 55.7	S56 47.5
03	336 39.4	255 08.9 ··	08.7	139 12.8 ··	55.4	127 32.0 ··	39.9	140 55.5 ··	01.5	Pollux	243 57.0	N28 04.2
04	351 41.8	270 08.1	09.0	154 14.1	56.0	142 34.3	39.9	155 57.8	01.6	Procyon	245 24.8	N 5 16.3
05	6 44.3	285 07.4	09.3	169 15.4	56.5	157 36.7	40.0	171 00.2	01.6			
06	21 46.8	300 06.6 N22	09.6	184 16.8 S 7	57.0	172 39.0 S10	40.0	186 02.6 S 4	01.7	Rasalhague	96 28.1	N12 34.4
07	36 49.2	315 05.8	09.9	199 18.1	57.5	187 41.4	40.1	201 05.0	01.7	Regulus	208 08.9	N12 03.4
08	51 51.7	330 05.1	10.1	214 19.4	58.0	202 43.7	40.1	216 07.3	01.8	Rigel	281 35.1	S 8 13.2
W 09	66 54.2	345 04.3 ··	10.4	229 20.7 ··	58.5	217 46.1 ··	40.2	231 09.7 ··	01.8	Rigil Kent.	140 23.9	S60 45.9
E 10	81 56.6	0 03.6	10.7	244 22.0	59.0	232 48.4	40.2	246 12.1	01.9	Sabik	102 39.4	S15 42.2
D 11	96 59.1	15 02.8	11.0	259 23.4	7 59.6	247 50.8	40.3	261 14.4	01.9			
N 12	112 01.5	30 02.1 N22	11.2	274 24.7 S 8	00.1	262 53.1 S10	40.3	276 16.8 S 4	02.0	Schedar	350 07.5	N56 26.1
E 13	127 04.0	45 01.3	11.5	289 26.0	00.6	277 55.5	40.4	291 19.2	02.0	Shaula	96 53.6	S37 05.5
S 14	142 06.5	60 00.6	11.8	304 27.3	01.1	292 57.8	40.4	306 21.5	02.1	Sirius	258 54.9	S16 41.5
D 15	157 08.9	74 59.8 ··	12.0	319 28.6 ··	01.6	308 00.2 ··	40.5	321 23.9 ··	02.1	Spica	158 56.2	S11 04.1
A 16	172 11.4	89 59.0	12.3	334 29.9	02.1	323 02.5	40.6	336 26.3	02.2	Suhail	223 10.3	S43 21.7
Y 17	187 13.9	104 58.3	12.6	349 31.3	02.6	338 04.9	40.6	351 28.7	02.2			
18	202 16.3	119 57.5 N22	12.9	4 32.6 S 8	03.2	353 07.2 S10	40.7	6 31.0 S 4	02.3	Vega	80 54.5	N38 46.1
19	217 18.8	134 56.8	13.1	19 33.9	03.7	8 09.6	40.7	21 33.4	02.3	Zuben'ubi	137 31.5	S15 58.1
20	232 21.3	149 56.0	13.4	34 35.2	04.2	23 11.9	40.8	36 35.8	02.4		S.H.A.	Mer. Pass.
21	247 23.7	164 55.3 ··	13.6	49 36.5 ··	04.7	38 14.3 ··	40.8	51 38.1 ··	02.4		° ′	h m
22	262 26.2	179 54.5	13.9	64 37.8	05.2	53 16.6	40.9	66 40.5	02.5	Venus	279 56.2	9 59
23	277 28.7	194 53.7	14.2	79 39.1	05.7	68 19.0	40.9	81 42.9	02.5	Mars	163 04.1	17 44
	h m									Jupiter	150 55.5	18 31
Mer. Pass.	4 37.1	v −0.7 d	0.3	v 1.3 d	0.5	v 2.4 d	0.1	v 2.4 d	0.0	Saturn	164 18.5	17 38

64

Appendix

STAR CHARTS

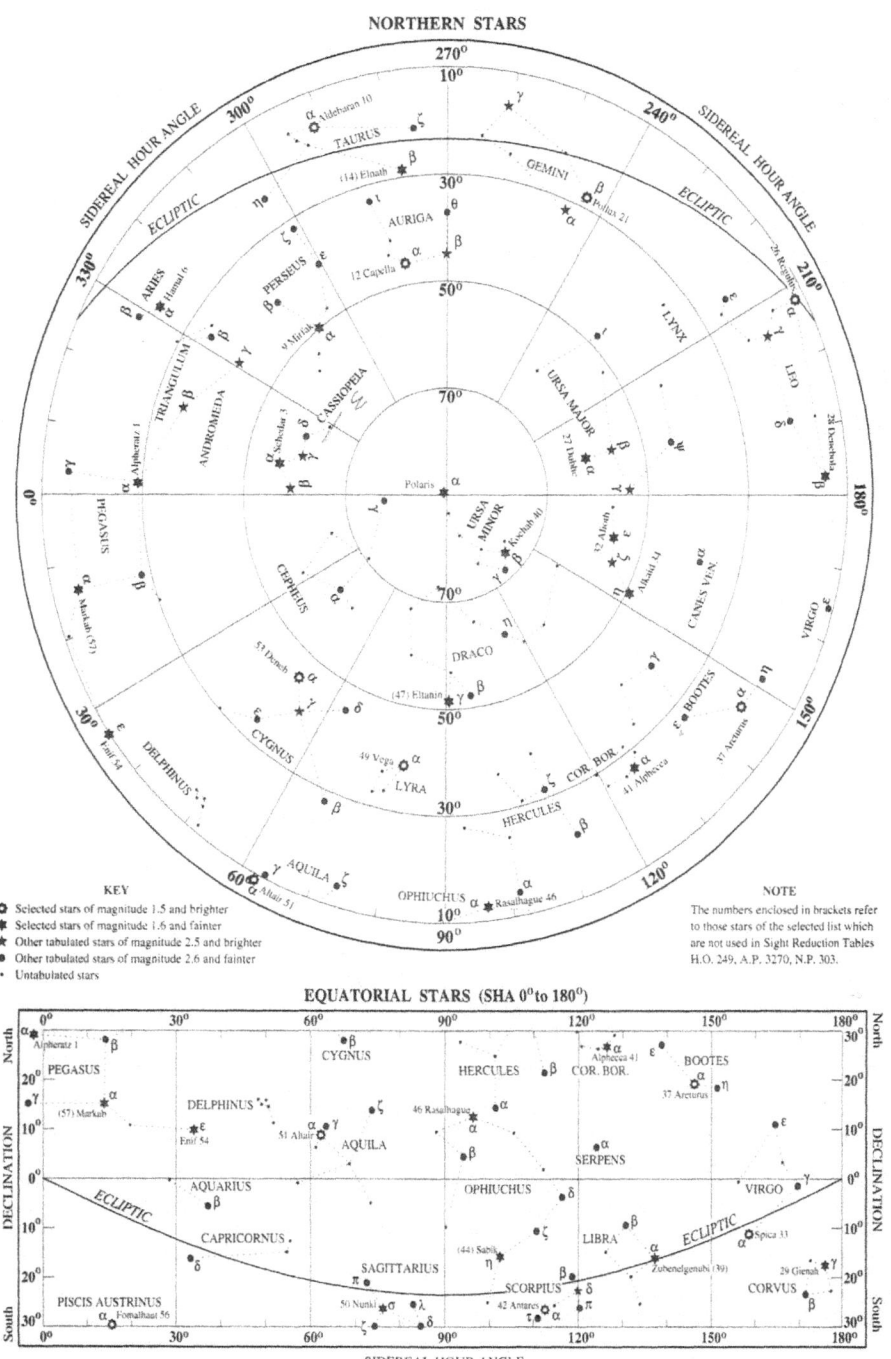

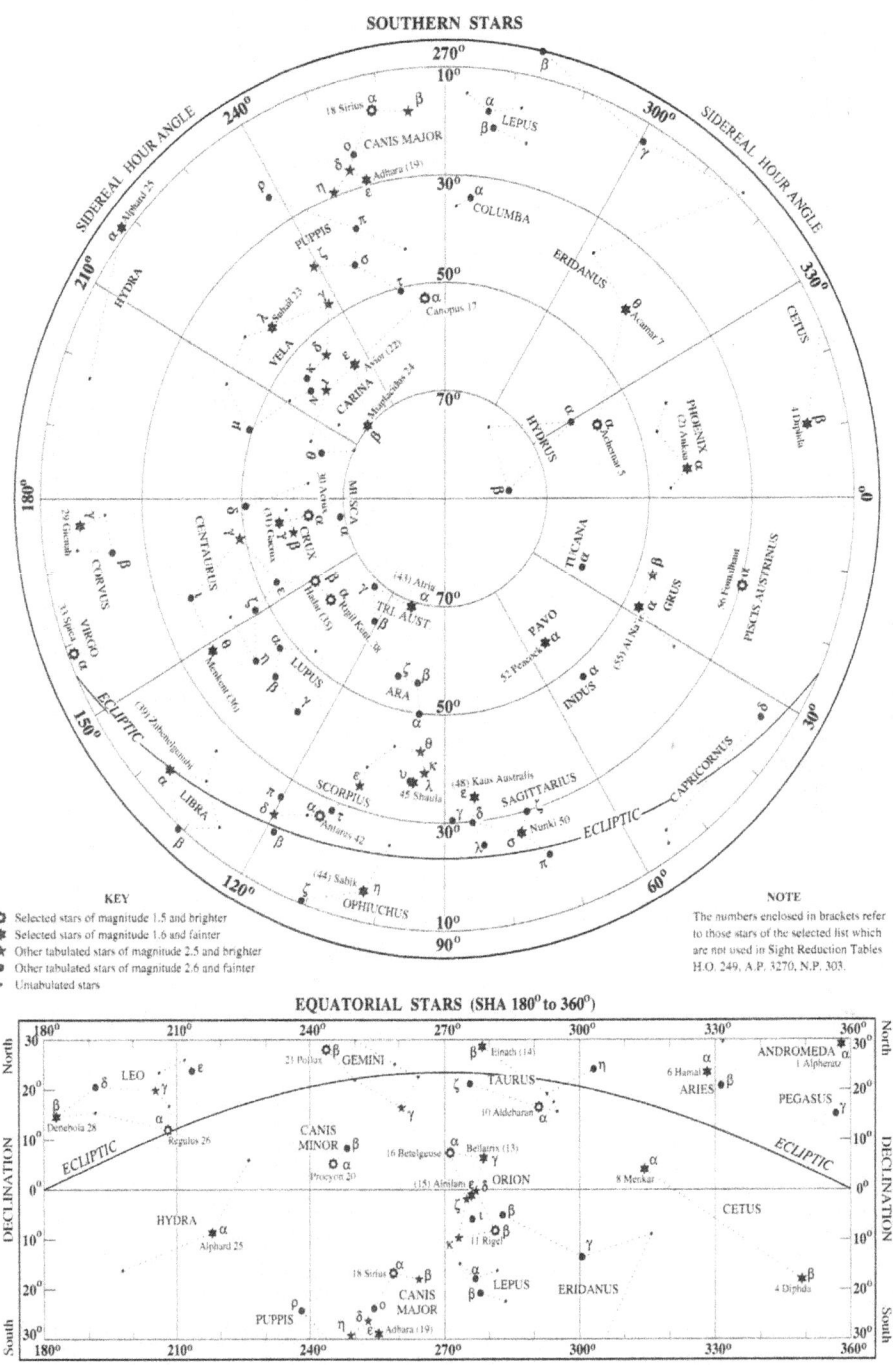

Appendix

Improving Template Readings

The templates are limited to latitude intervals of 10° in multiple of 15°, i.e., there is a 35° template and a 45° template. Best results occur for observations at a latitude that matches a template; between those latitudes, the results are a bit less accurate, as was illustrated in Table 5-2. It is not often we need higher precision than this system provides, but if so there is a way to enhance the readings.

At 40 N, for example, we are right in the middle of two templates, and thus expect the worst errors, as shown in Table 5-2. In the procedure below we can manually offset the template to get better results.

Step 1. Draw in the observer's meridian line from the active LHA Aries to the center of the centerpin as discussed in Section 7.2

Step 2. With the blue template removed from the centerpin, position it so the blue arrowed line remains over the observer's meridian and slide it up or down until the observer's true latitude is centered on top of the center pin as shown in Figure 7-4.

Step 3. Hold the disk in place to read the new values of the Hc and Zn of the stars or planets in question.

In this example with true Lat = 40 10' N and LHA Aries of 344.4, we should read what was listed in Table 5-2.

Capella	24° @ 049
Fomalhaut	21° @ 181
Vega	40° @ 292

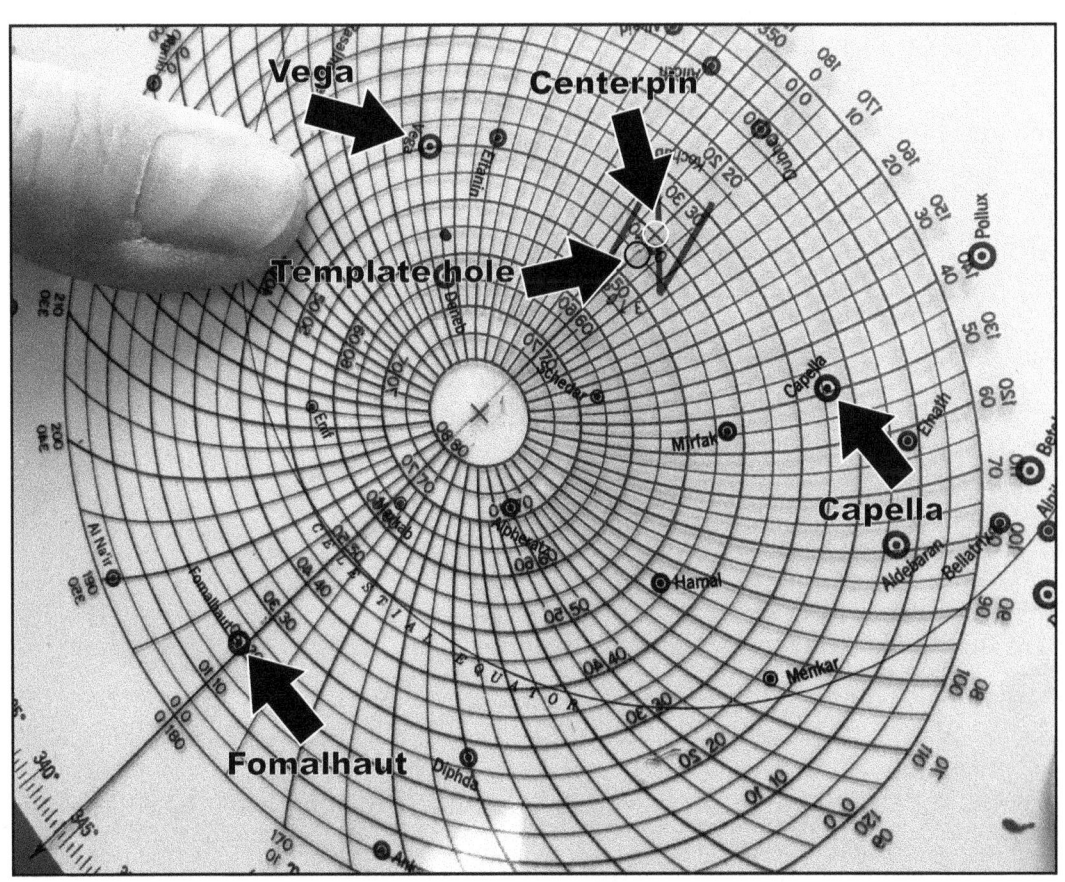

Reading Hc and Zn from an template offset to match the true latitude.

Equation Summary

UTC = LMT + DR-Lon(W)
UTC = LMT − DR-Lon(E)

Rim Scale = LHA Aries

Rim Scale = 360° − SHA

GHA Star = GHA Aries + SHA star

LHA Aries = GHA Aries − DR-Lon(W)
LHA Aries = GHA Aries + DR-Lon(E)

SHA = 360' − Rim Scale

Computed Star ID

www.starpath.com/usno
www.starpilotllc.com
www.stellarium.org

www.ingramcontent.com/pod-product-compliance
Lightning Source LLC
Chambersburg PA
CBHW080525110426
42742CB00017B/3241